Useful Physical Constants

Boltzmann's constant, k	8.62×10^{-5} ev/°K
Electron rest mass, m_0	9.11×10^{-28} g
Energy associated with 1 eV	1.6×10^{-12} erg
Magnitude of the electronic charge, q	1.6×10^{-19} coulomb
Permittivity of free space, ϵ_0	8.86×10^{-14} farad/cm
Planck's constant, h	4.14×10^{-15} eV-sec
Room temperature value of kT	0.0259 eV
Velocity of light, c	3×10^{10} cm/sec
Angstrom unit, Å	10^{-8} cm
Micron, μm	10^{-4} cm
Thousandth of an inch, mil	25.4 μm

Semiconductor Power Devices

Thyristor devices. Courtesy of the General Electric Company, Semiconductor Products Dept., Auburn, N.Y.

SEMICONDUCTOR POWER DEVICES

PHYSICS OF OPERATION AND FABRICATION TECHNOLOGY

SORAB K. GHANDHI

Rensselaer Polytechnic Institute

A WILEY-INTERSCIENCE PUBLICATION

JOHN WILEY & SONS, New York • London • Sydney • Toronto

Library of Congress Cataloging in Publication Data:

Ghandhi, Sorab Khushro, 1928-
    Semiconductor power devices.

    "A Wiley-Interscience publication."
    Includes bibliographical references and index.
    1.   Power semiconductors.   I.   Title.

TK7871.85.G43   621.31′7   77-8019
ISBN 0-471-02999-8

Printed in the United States of America

10 9 8 7 6 5 4 3 2 1

To

MY FATHER

in memoriam

PREFACE

The use of solid state techniques for electrical power control coincides with the development of large area, single crystal rectifiers in the early 1950s, followed by the development of power transistors. Today such devices have an important part in this field, with silicon being the almost universal choice of material. A true revolution did not take place, however, until the announcement of the *p-n-p-n* device* by the Bell Laboratories. This device was originally developed for use as a cross-point switch in telephone networks. The General Electric Company was the first to exploit its potential in high power control applications, with the successful demonstration of a silicon device capable of operation at high current levels. Originally called the Silicon Controlled Rectifier, devices of this type (together with a number of derivative structures) are now grouped under the generic name of *thyristor*, and are the most important members of the family of semiconductor power devices. They have found their way into applications ranging from speed control in home appliances to switching and power inversion in high voltage d-c transmission lines.

For a long time, semiconductor power devices (and thyristors in particular) were not considered part of the mainstream of solid state endeavor, which was dominated by the area of microelectronics. Research in the field was confined to a small group of workers, and reported in the specialist literature. The first attempt to integrate this material was made in 1964 in a book† that described the principles and circuit applications of thyristor devices. Since that time, most new books have been restricted to circuit applications of these devices.

The present volume is concerned with the physics of operation and the fabrication technology of power diodes, transistors, and thyristors. It has

---

*J. L. Moll, M. Tannenbaum, J. M. Goldey, and N. Holonyak Jr., "*p-n-p-n* Transistor Switches", *Proc. IRE*, **44**, No. 9, pp. 1174–1182 (1956).

†F. E. Gentry, F. W. Gutzwiller, N. Holonyak, Jr., and E. E. Von Zastrow, *Semiconductor Controlled Rectifiers*, Prentice-Hall, Englewood Cliffs, N.J. (1964).

been written to meet the needs of a rapidly increasing body of engineers who are becoming involved in the field of power control. This includes engineers who must design advanced circuits using these devices, and thus need to better understand the physics of device behavior, as well as those who must specify device characteristics that are necessary for the new circuit applications. It will also meet the needs of those who must design and fabricate these new devices.

Finally, this book has been written to provide a text that can bring this material into the academic curriculum. Many students are interested in this area because of its growing importance to industry. Unfortunately, university courses in semiconductor power devices have been initiated only recently, since the required information has been scattered through the technical literature. This book should give the necessary impetus for such course offerings—it coordinates and unifies this material into a logical form.

The book is directed at readers who already are familiar with the physics of operation and the fabrication technology of transistors and microcircuits. This has been done to avoid repeating material already adequately presented in a number of books and to keep the book to a reasonable size. I have attempted to emphasize areas that are unique and of special importance to semiconductor power devices, organizing material so as to bring out these areas in some detail.

Chapter 1 deals with such fundamental considerations as lifetime, with emphasis placed on high level effects and on Auger processes. Also included is a discussion of transport phenomena in high resistivity materials, and the development of ambipolar parameters for their characterization. A major section of this chapter deals with single and double injection, which leads to the formation of microplasmas and mesoplasmas with eventual destruction by second breakdown.

Chapter 2 deals with the characteristics of diodes with high reverse breakdown voltage. Techniques for obtaining these characteristics are developed. Included are discussions of the use of guard rings, field plates, and surface contouring methods, all of which are unique to power devices. Instabilities that can be encountered during operation in this mode are also described.

Chapter 3 is about the forward-biased diode. Here device operation at high injection levels is considered, together with the special problems of highly doped emitters. The main emphasis is on the $p^+\text{-}i\text{-}n^+$ structure which closely represents the behavior of a thyristor in its ON state. Both static and dynamic characteristics of this device are considered in some detail. Thermal instabilities in forward-biased diodes are also examined.

Chapter 4 deals with the special characteristics of high voltage, high power transistors. These include double saturation effects in their output characteristics, as well as falloff in current gain and in gain-bandwidth product at high injection levels. An important part of this chapter treats phenomena leading to the formation of mesoplasmas and second breakdown, as well as corrective measures for alleviating these effects.

Chapter 5 discusses thyristors and related devices such as gate-controlled switches, diacs, and triacs. Topics include a study of various modes of device operation, such as reverse blocking, forward blocking, and forward conduction. Dynamic processes are considered together with the special problems of turn-on and reverse recovery. Techniques involving the use of shorts and amplifying gates are covered in detail. Additional topics include light-activated thyristors, amplifying gate thyristors, and reverse conducting thyristors, as well as gate-assisted turn-off devices. Mechanisms for device instability are also outlined.

Chapter 6 deals with fabrication technology. The emphasis again is on aspects that are unique to power devices, since it is assumed that the reader has some familiarity with the basic materials and processing problems and techniques encountered with lower power devices and microcircuits.* Topics such as neutron doping, deep diffusions, stress-free processing, the control of lifetime, and special packaging problems for these devices are emphasized. Hard and soft solder systems are considered for device mounting, as is the use of intermediate materials for the reduction of interfacial stress during thermal cycling. Finally, integrated circuit packaging techniques are described, since these present an important area for control applications at medium power levels.

I have attempted to be as comprehensive as possible. However, progress in this field is very rapid, and any attempt to be encyclopaedic in coverage would only result in early obsolescence. Consequently, the emphasis is on the basic principles of device operation to enable the reader to understand and evaluate new developments in the field as they are announced. With this in mind, no attempt has been made to use references as a means for giving credit to the persons who did the original research; rather, references have been chosen to provide a useful mechanism for further study. In like manner, a number of problems have been provided, many of which were selected from practical situations and are intended to bring out points not covered in detail in the text.

I am indebted to many people who have contributed, directly or indirectly, to the preparation of this book. Many thanks are due to Dr. A. P. Ferro and Dr. D. Schaefer, as well as to R. Guess, all of the General Electric Corporate Research and Development Center, for providing the

---

*See, for example, S. K. Ghandhi, *The Theory and Practice of Microelectronics*, John Wiley and Sons, New York, 1968.

stimulus for a power semiconductor program at Rensselaer. I am also indebted to the faculty members in our solid state program, with whom I have had many fruitful discussions. These include Drs. J. M. Borrego, B. K. Bose,* P. K. Das, R. J. Gutmann, K. Rose, and A. J. Steckl. In addition, technical discussions with many friends at the General Electric Company, which took place during the presentation of a series of lectures on this subject, are gratefully acknowledged. I must single out Dr. B. Jayant Baliga of the Power Semiconductor Branch, General Electric Corporate Research and Development Center, for his friendly but critical review of the manuscript and for his detailed, constructive suggestions. His interest has been very helpful and stimulating and has greatly enhanced the relevance of this work to modern practice in the field.

The material of this book has been taught, in one form or another, to graduate students over the last four years. They deserve my heartfelt thanks and sympathy. Their penetrating questions have often led to rethinking and reworking of this text over the years. Finally, credit for the typing of this manuscript and its many revisions must go to R. Carla Rafun, who also played a major role in editing and checking the manuscript, right from its typed version up to the page proof stage. I am sure this manuscript would never have been completed were it not for her selfless and enthusiastic participation in what could easily have been considered just one more item of overwork.

SORAB K. GHANDHI

Niskayuna, New York
June 1977

*Now with the General Electric Research and Development Center, Schenectady, N.Y.

CONTENTS

Semiconductor Power Devices

CHAPTER

1

Fundamental Considerations

CONTENTS

AT FIRST GLANCE, it would appear that the physics of semiconductor power devices is identical to that of their low power counterparts. Certainly device operation is the same, regardless of physical size. Nevertheless, many aspects of device physics take on a different sense of importance under conditions of high power operation. Furthermore, certain device structures, such as *p-i-n* diodes and *p-n-p-n* thyristors, have unique advantages for high power applications and are used primarily in these situations alone.

This chapter outlines some fundamental considerations that arise in a study of semiconductor power devices. A priori, such devices must be capable of dissipating large amounts of power, under both steady state and transient conditions. Thus the problems of heat removal represent an important aspect of fabrication and packaging technology. In addition, the devices must be operated in a switching mode. In such operation, the power handling capacity of a device is directly related to its ability to support high reverse voltages in the OFF state, and to conduct at high forward current levels in the ON state.

The reverse breakdown voltage of a junction diode is determined by the resistivity of the *p-* and *n-*regions, and by its doping profile. This problem is considered in detail in Chapter 2. Qualitatively, however, it is seen that the ability to hold off high reverse voltages mandates the use of materials with high starting resistivity. Values as high as 500 ohm-cm are encountered in diodes designed to support reverse potentials in the kilovolt range. By way of comparison, the starting resistivity of devices used in digital integrated circuits is typically 0.15–0.2 ohm-cm!

The ability to handle large forward currents necessitates device operation at high levels of injected carrier concentration, often well in excess of the initial concentration of thermally ionized carriers present in the semiconductor. Thus the background concentration of the semiconductor becomes relatively unimportant in determining the device behavior. Furthermore, the total electron and hole concentrations become approximately equal in magnitude, so that injection at high levels shares many common characteristics with injection into intrinsic materials. It is appropriate, therefore, to give special consideration to transport phenomena in materials that exhibit intrinsic conduction, especially in lightly doped semiconductors. For silicon, material with carrier concentrations up to $10^{15}/\text{cm}^3$ is classed as high resistivity material. In high voltage devices, the initial carrier concentration is often as low as $10^{13}/\text{cm}$.

1.1 LIFETIME

A semiconductor is said to be in *thermal equilibrium* with its environment if for any process, there is an inverse process that occurs at the same rate.

At any given temperature, this results in the continual generation and recombination of hole-electron pairs with identical generation and recombination rates. The end result is an equilibrium concentration of holes $\bar{p}$ and electrons $\bar{n}$ such that $\bar{p}\bar{n} = n_i^2 = $ constant. If such a system is excited (e.g., by illumination with steady penetrating light), both carrier generation and recombination rates will be altered until new steady state conditions are reached. At this point, the electron and hole concentrations are given by

$$n = \bar{n} + n' \tag{1.1a}$$

$$p = \bar{p} + p' \tag{1.1b}$$

where n', p' are the time-dependent concentrations of excess electrons and holes, respectively, and $\bar{n}, \bar{p}$ are their thermal equilibrium values. Upon removal of this stimulus, a process of recombination occurs, until the system returns to its thermal equilibrium value. In an n-type semiconductor, the recombination process for minority carriers can be approximated by a power series, such that

$$-\frac{\partial p'}{\partial t} = \gamma_1 p' + \gamma_2 p'^2 + \gamma_3 p'^3 + \cdots \tag{1.2}$$

Here the different terms can be expected to dominate over different regimes of carrier concentration. From a physical point of view, the second-order term is characteristic of *band-to-band recombination*, accompanied by photon emission. Radiative processes of this type are highly improbable in indirect gap materials such as silicon, and recombination is more appropriately described by

$$-\frac{\partial p'}{\partial t} = \gamma_1 p' + \gamma_3 p'^3 \tag{1.3}$$

for this material. For a very wide range of injection levels, the first term predominates, and is useful in characterizing *phonon-assisted recombination*, via deep impurity levels. The parameter $1/\gamma_1$ is referred to as the *minority carrier lifetime* τ_p. The evaluation of this term has been extensively treated in the literature [1, 2] and is reviewed here for the case of a single recombination center. The situation for multiple levels is extremely complex [3] and is often treated in terms of the single level case.

Consider an n-type semiconductor, with a density of recombination centers N_r at an energy level E_r. For the purposes of discussion it is assumed that these centers are donors. In thermal equilibrium, some of these are neutral and the rest ionized. Thus $N_r = \bar{N}_r^+ + \bar{N}_r^0$, where the bar indicates thermal equilibrium values. Under *steady state nonequilibrium* conditions, the concentrations of neutral and ionized recombination centers are altered, and we have $N_r = N_r^+ + N_r^0$.

The rate of capture of electrons from the conduction band is proportional to the electron concentration and to the concentration of positively ionized centers. Thus

$$R_{cn} = \alpha_n n N_r^+ \tag{1.4a}$$

where α_n is a rate constant, given by the product of the electron capture cross section and the electron thermal velocity.

The rate of emission of electrons to the conduction band is proportional to the concentration of neutral centers and to the density of available states at the conduction band edge. Thus

$$R_{en} = C_1 N_r^0 (N_c - n) \cong C_1 N_r^0 N_c \tag{1.4b}$$

where C_1 is a proportionality factor, and N_c is the density of states at the conduction band edge.

In like manner, the capture and emission rates of holes with respect to the valence band are given by

$$R_{cp} = \alpha_p p N_r^0 \tag{1.5a}$$

$$R_{ep} = C_2 N_r^+ N_v \tag{1.5b}$$

where C_2 is a proportionality factor, and N_v is the density of states at the valence band edge.

The factors C_1 and C_2 may be evaluated by using the conditions for thermal equilibrium (i.e., $\bar{R}_{cn} = \bar{R}_{en}$ and $\bar{R}_{cp} = \bar{R}_{ep}$). Making these substitutions into (1.4a) and (1.4b) and solving, gives

$$R_{cn} - R_{en} = -\frac{dn'}{dt} = \alpha_n \left(n N_r^+ - n_1 N_r^0 \right) \tag{1.6}$$

where

$$n_1 = N_c \exp - \left(\frac{E_c - E_r}{kT} \right) \tag{1.7}$$

In like manner, the rate of change of the excess hole concentration is given by

$$R_{cp} - R_{ep} = -\frac{dp'}{dt} = \alpha_p \left(p N_r^0 - p_1 N_r^+ \right) \tag{1.8}$$

where

$$p_1 = N_v \exp - \left(\frac{E_r - E_v}{kT} \right) \tag{1.9}$$

In these equations, n_1, p_1 can be thought of* as the electron and hole concentrations if the Fermi level were at the recombination level; k is the Boltzmann constant, and T is the temperature in degrees Kelvin.

In steady state, electrons and holes are simultaneously annihilated, so that

$$\frac{dn'}{dt} = \frac{dp'}{dt} \tag{1.10}$$

In addition,

$$N_r = N_r^0 + N_r^+ \tag{1.11}$$

Combining (1.6) to (1.11), the rate of recombination in steady state may be written as $-U$, where

$$-U = \frac{dn'}{dt} = \frac{dp'}{dt} = \frac{np - n_i^2}{\tau_{p0}(n + n_1) + \tau_{n0}(p + p_1)} \tag{1.12}$$

and

$$\tau_{p0} = \frac{1}{\alpha_p N_r} \tag{1.13a}$$

$$\tau_{n0} = \frac{1}{\alpha_n N_r} \tag{1.13b}$$

Finally, since a first-order recombination process is assumed, the minority carrier lifetime is given from (1.3) by $-p'/(dp'/dt) = -p'/U$.

Equation 1.12 can be applied to a number of special cases, which we now consider.

In the *space charge neutral* case, $n' = p'$ and the lifetime is given by

$$-\frac{n'}{U} = -\frac{p'}{U} = \tau_{p0} \left(\frac{\bar{n} + n_1 + n'}{\bar{n} + \bar{p} + n'} \right) + \tau_{n0} \left(\frac{\bar{p} + p_1 + n'}{\bar{n} + \bar{p} + n'} \right) \tag{1.14}$$

---

*These are fictitious quantities, introduced in the interest of compactness.

For n-type material, $\bar{p} \ll \bar{n}$. Writing $h = n'/\bar{n}$,

$$\tau_p = \tau_{p0}\left[1 + \frac{1}{1+h}\exp\left(\frac{E_r - E_f}{kT}\right)\right]$$

$$+ \tau_{n0}\left[\frac{h}{1+h} + \frac{1}{1+h}\exp\left(\frac{E_i - E_r}{kT} + \frac{E_i - E_f}{kT}\right)\right] \qquad (1.15)$$

In deriving this expression it is assumed that $N_c = N_v$ and $E_i = (E_c + E_v)/2$. For *low level injection* in an n-type semiconductor, $h \ll 1$, so that

$$\tau_{p,\text{low}} = \tau_{p0}\left[1 + \exp\left(\frac{E_r - E_f}{kT}\right)\right] + \tau_{n0}\exp\left[\left(\frac{E_i - E_r}{kT}\right) + \left(\frac{E_i - E_f}{kT}\right)\right] \quad (1.16)$$

For *high level injection*, $h \gg 1$, and

$$\tau_{p,\text{high}} = \tau_{p0} + \tau_{n0} \qquad (1.17)$$

Note that the lifetime for high injection levels is independent of the doping level. This can be expected, since the injected carrier concentration for this situation is well in excess of the background concentration.

Conditions in the depletion layer can be approximated by setting $n \cong p \cong 0$, so that the *generation* rate is given by

$$U = \frac{n_i^2}{n_1\tau_{p0} + p_1\tau_{n0}} \equiv \frac{n_i}{2\tau_{sc}} \qquad (1.18)$$

where τ_{sc} is defined as the *space charge generation lifetime*. Solving,

$$\tau_{sc} = \frac{\tau_{p0}}{2}\left[\exp\left(\frac{E_r - E_i}{kT}\right)\right] + \frac{\tau_{n0}}{2}\left[\exp\left(\frac{E_i - E_r}{kT}\right)\right] \qquad (1.19)$$

This lifetime is also independent of the doping level, as can be expected for a fully depleted region of a semiconductor. Note that it is equal in magnitude to the low level lifetime in an intrinsic semiconductor.

The behavior of the low level and space charge generation lifetimes can be conveniently combined in a single display for any specific ratio of τ_{n0}/τ_{p0}. Figure 1.1, for example, shows the appropriate minority carrier lifetime for n- and p-type semiconductors for various positions of the Fermi level. The precise nature of this curve depends on the location of the

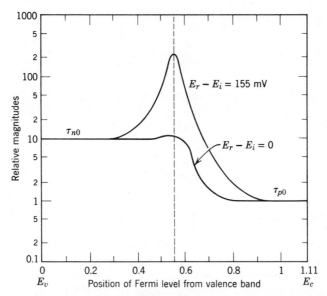

Fig. 1.1   The variation of lifetime with position of the Fermi level.

recombination level with respect to the intrinsic level and on the values of τ_{n0} and τ_{p0}. One typical case, in which there is a finite separation* between the recombination level and the intrinsic level (i.e., for $E_r - E_i = 155$ mV) is shown, together with the special case of a mid-band recombination center ($E_r = E_i$) for which the minority carrier lifetime varies nearly monotonically from τ_{n0} to τ_{p0}. A value of $\tau_{n0} = 10\tau_{p0}$ was chosen in this figure because it represents a practical situation.

Experimental verification of the theory of recombination via deep impurity levels has been obtained [4] by a method based on the measurement of the decay time of the photocurrent, generated by pulsed irradiation of the sample with electrons. The theory is found to be quite well suited for $0.01 \leqslant h \leqslant 100$; that is, for injection levels ranging from 2 decades below the background concentration to 2 decades above it.

Additional recombination processes have been observed [5, 6] in indirect gap semiconductors when one or more mobile species are in excess. For example, if a large concentration of electrons is present, it is possible to obtain direct recombination between an electron and a hole, accompanied by the transfer of energy to another free electron. Such a process, involving two electrons and one hole, occurs when holes are injected into a heavily doped n^+-region, as in the emitter of an n^+-p-n transistor. It results in

*Both positive and negative values of $E_r - E_i$ occur in practice.

extremely fast recombination, characterized by

$$-\frac{dp'}{dt} = \gamma_{3p} n^2 p' \qquad (1.20a)$$

where $\gamma_{3p} = 1.7 \times 10^{-31}$ cm^6/sec for silicon.

In like manner, recombination in a heavily doped p^+-region can occur by involving two holes and one electron, so that

$$-\frac{dn'}{dt} = \gamma_{3n} p^2 n' \qquad (1.20b)$$

where $\gamma_{3n} = 1.2 \times 10^{-31}$ cm^6/sec for silicon. These processes are known as Auger processes.

High level injection in the mid-region of p-i-n diode, as well as in the wide base region of a p-n-p-n thyristor, is also dominated by this type of recombination process. This is especially true under surge conditions, where current densities are encountered in the 1000–3000 A/cm^2 range and injected carrier concentrations can be 3 to 4 decades higher than the background concentration. Here, however, large numbers of electrons and holes are involved, in equal number, thus both process occur. For n-type silicon, recombination is characterized by

$$-\frac{dp'}{dt} = \gamma_3 p'^3 \qquad (1.20c)$$

where $\gamma_3 = 2$–2.9×10^{-31} cm^6/sec.

In all the foregoing situations, the Auger processes can be described by a lifetime τ_A, equal to the ratio of the excess carrier concentration to the rate of decay of excess carriers. The effective lifetime for an n-type indirect gap semiconductor, which includes both phonon-assisted recombination and Auger recombination, follows from (1.3) as

$$\frac{1}{\tau_{\text{eff}}} = \frac{1}{\tau_p} + \frac{1}{\tau_A} \qquad (1.21)$$

where τ_p is the minority carrier lifetime. Moreover, τ_p is the high level lifetime when large concentrations of both electrons and holes are involved. However the low level lifetime must be used when only one mobile species is present in high concentrations.

1.2 CARRIER TRANSPORT

For the quantitative study of carrier transport in lightly doped semiconductors, consider a semiconductor in which recombination occurs via deep impurity levels. The continuity and diffusion equations for an elemental volume of this material may be written as

$$\frac{\partial n'}{\partial t} = U + D_n \frac{\partial^2 n'}{\partial x^2} + \mu_n \frac{\partial}{\partial x}(n\mathcal{E}) \tag{1.22a}$$

$$\frac{\partial p'}{\partial t} = U + D_p \frac{\partial^2 p'}{\partial x^2} + \mu_p \frac{\partial}{\partial x}(p\mathcal{E}) \tag{1.22b}$$

Here U is the net generation rate; D_p, D_n are the hole and electron diffusion constants; μ_p, μ_n are the hole and electron mobilities; and $\mathcal{E}$ is the electric field. In addition, charge neutrality holds approximately (even in the presence of an electric field), so that $\partial n' \cong \partial p'$.

Multiplying (1.22a) by $\mu_p p$, and (1.22b) by $\mu_n n$, and combining with the Einstein relation, $D = (kT/q)\mu$, we have

$$\frac{\partial p'}{\partial t} = -\frac{p'}{\tau_a} + D_a \frac{\partial^2 p'}{\partial x^2} - \frac{n-p}{n/\mu_p + p/\mu_n} \mathcal{E} \frac{\partial p'}{\partial x} \tag{1.23}$$

where

$$D_a = \frac{n+p}{n/D_p + p/D_n} \tag{1.24}$$

$$\tau_a = \frac{-p'}{U} = \frac{-n'}{U} \tag{1.25}$$

and D_a and τ_a are known as the *ambipolar diffusion constant* and the *ambipolar lifetime*, respectively. In most situations, the entire term in $\mathcal{E}$ can be neglected as second order in comparison to the others. In like manner, the continuity and diffusion equation for minority carriers in a p-type semiconductor can be written as

$$\frac{\partial n'}{\partial t} = -\frac{n'}{\tau_a} + D_a \frac{\partial^2 n'}{\partial x^2} \tag{1.26}$$

The use of ambipolar parameters casts the coupled sets of differential equations (1.22a) and (1.22b) in the form of independent equations, and allows the study of high level behavior in terms of well-established low level theories. However D_a is a function of the injection level in this equation. In the limit of high level injection, the ambipolar diffusion constant is independent of carrier concentration and is given by

$$D_a = \frac{2D_n D_p}{D_n + D_p} = \frac{2D_n}{1+b} \tag{1.27}$$

where $b = \mu_n / \mu_p$. Figure 1.2 represents the behavior of μ_n and μ_p as a function of the impurity concentration. Their temperature dependence is shown in Fig. 1.3. For silicon, $b = 3$, so that

$$D_a = 1.5 D_p = 0.5 D_n \tag{1.28}$$

The ambipolar lifetime can also be determined for high level injection conditions, where $n' \gg \bar{n}$ and $p' \gg \bar{p}$. Making these substitutions into (1.12), and noting that $n' = p'$, gives

$$-U = \frac{n'}{\tau_{p0} + \tau_{n0}} = \frac{p'}{\tau_{p0} + \tau_{n0}} \tag{1.29}$$

Thus for high level injection conditions, the ambipolar lifetime is given by

$$\tau_a = \tau_{p0} + \tau_{n0} \tag{1.30}$$

which is the high level lifetime.

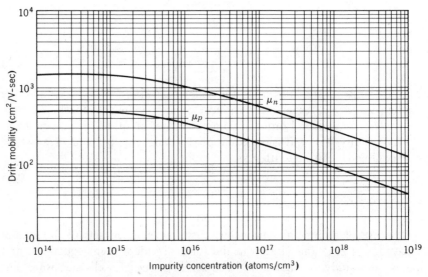

Fig. 1.2 Mobility versus impurity concentration.

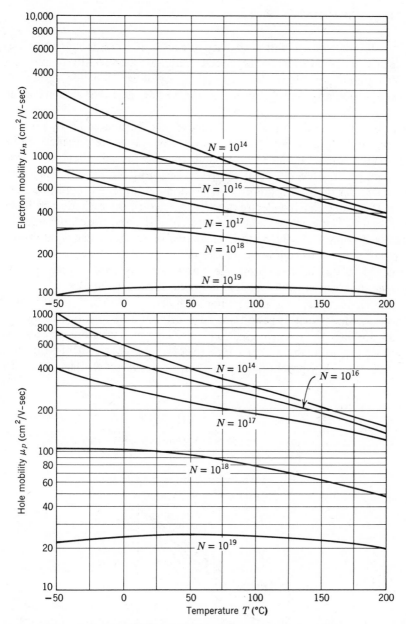

Fig. 1.3 Mobility versus temperature. From *Transistor Engineering* by A. Phillips. Copyright 1962. Used with permission of McGraw-Hill Book Company.

1.3 SINGLE INJECTION

Semiconducting silicon, with resistivity in excess of 10 ohm-cm (*n*-type), is generally considered to be lightly doped. Materials with resistivity in excess of 100 ohm-cm are often classed as *semi-insulating*. Depleted regions in semiconductors are semi-insulating in their characteristics and are often classed as such.

Injection in a lightly doped semiconductor can be treated [7] by initially assuming that it is a perfect insulator, characterized by the presence of one deep-lying recombination level. Assume for the present that the Fermi level is above this level, so that the recombination centers are filled with electrons under thermal equilibrium conditions. A voltage is applied across the bar (Fig. 1.4), causing electrons to be injected at the cathode and holes at the anode. The electrons have infinite lifetime since the recombination centers are completely filled (thus incapable of further electron capture), and move toward the anode under the influence of the electric field. The lifetime of holes injected from the anode side is short, since all the recombination centers are available for their attractive capture.

If the bar is "long," the transit time of holes can be much greater than the hole lifetime. For this condition, holes are not able to traverse the bar, thus do not alter the injected electron space charge. Therefore, injection is essentially due to a single carrier type (electrons) and is controlled by this space charge. This process is known as *single injection*. The transport

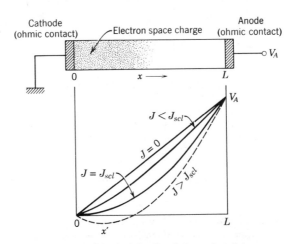

Fig. 1.4 Single Injection into an Insulator.

equations for this process are

$$\frac{d^2V}{dx^2} = -\frac{d\mathscr{E}}{dx} = -\frac{\rho}{\varepsilon\varepsilon_0} = \frac{qn}{\varepsilon\varepsilon_0} \tag{1.31}$$

and

$$J = -nq\mu\mathscr{E} \tag{1.32}$$

Here V is the voltage, J is the current density, ρ is the charge density, q is the magnitude of the electronic charge, n is the electron concentration, $\mathscr{E}$ is the electric field, ε is the relative permittivity, ε_0 is the permittivity of free space, and μ is the mobility. Solving

$$\mathscr{E}^2 = \left(\frac{dV}{dx}\right)^2 = \left(\frac{2J}{\varepsilon\varepsilon_0\mu}\right)x + \mathscr{E}_C^2 \tag{1.33}$$

where $\mathscr{E}_C$ is the electric field at the cathode. For a constant applied voltage, the value of $\mathscr{E}_C$ depends on the current density.

Solutions of voltage as a function of distance along the bar are shown in Fig. 1.4 for different values of J. From these solutions we note that for $J = 0$, the V versus x curve is a straight line (i.e., constant field throughout the bar). With increasing J, the V versus x curve becomes depressed (i.e., the field at the cathode falls). At $J = J_{scl}$, the field becomes zero at the cathode. At this point, the forward current is completely space charge limited, and J_{scl} is defined as the *space charge limited current density*.

The situation for $J > J_{scl}$ is interesting because it results in the formation of a virtual cathode at $x = x'$, where the electric field is zero. However our attention is confined to the space charge limited case. For this case, solution of (1.33) gives

$$J_{scl} = \frac{9}{8}\frac{\varepsilon\varepsilon_0\mu}{L^3}V_A^2 \tag{1.34}$$

where L is the length of the bar and V_A is the applied voltage.

We can now include the effect of background carrier concentration N_B. Initially, current flow must be ohmic, so that

$$J = \frac{q\mu N_B V_A}{L} \tag{1.35}$$

Eventually, ohmic behavior changes to space charge limited behavior when

the injected carrier concentration becomes equal to the background concentration. This occurs at a voltage V_1, obtained by equating (1.34) and
(1.35), so that

$$V_1 = \frac{8}{9} \frac{q N_B L^2}{\varepsilon \varepsilon_0} \tag{1.36}$$

1.3.1 Partially Filled Recombination Centers

When some of the recombination centers are initially empty, the lifetime is
finite, since some electrons are lost in filling up these centers. Initially the
current is reduced by a factor $K = n/(n + n_r)$, where n_r is the concentration
of empty recombination centers. Let N_r be the total number of recombination centers, located at E_r. Then

$$n_r = N_r \exp\left[\frac{-(E_r - E_f)}{kT} \right] \tag{1.37}$$

where E_f is the Fermi level, k is the Boltzmann constant, and T is the
temperature (°K). If N_c is the density of states at the conduction band
edge,

$$n = N_c \exp\left[\frac{-(E_c - E_f)}{kT} \right] \tag{1.38}$$

so that

$$K = N_c \Bigg/ \left[N_c + N_r \exp\left(\frac{-(E_c - E_r)}{kT} \right) \right] \tag{1.39}$$

and the initial current is reduced by this factor. The transition voltage to
space charge limited current flow is correspondingly altered by this same
factor, to a value V_1'.

With increasing electron concentration, a point is reached at which all
recombination centers are filled. At this voltage V_2, the current abruptly
returns to its original value, given by (1.34).

The charge required to fill the recombination centers is $q n_r L$ per unit
cross-sectional area. This charge must be supported by the voltage V_2,

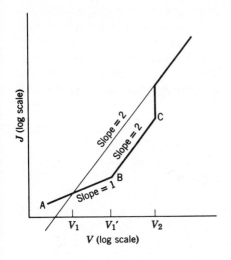

Fig. 1.5 *V-I* **Characteristics for single injection.**

across a bar capacitance of $\varepsilon\varepsilon_0/L$ per unit area. Then

$$V_2 = \frac{qn_r L}{\varepsilon\varepsilon_0/L}$$

$$= \frac{qn_r L^2}{\varepsilon\varepsilon_0} \tag{1.40}$$

Figure 1.5 shows the idealized $V-J$ characteristic describing single carrier injection, on the assumption that $V_2 > V_1'$. The region BC would not exist if $V_2 \leqslant V_1'$.

1.3.2 Punchthrough

The phenomenon of punchthrough in a transistor can be explained in terms of single carrier injection into insulators. Consider, for example, an n^+-p-n transistor in the common emitter configuration, with a supply voltage of $+V_{CC}$, relative to the emitter. Here the collector-base depletion layer penetrates into the base region and eventually reaches the emitter at a sufficiently high collector voltage. This condition is known as *punchthrough*, or *reachthrough*. At this point, the n^+-emitter can inject electrons into a depleted base region, with injected electrons flowing to the positive contact at the collector.* Device behavior is thus one of single

---

*On the other hand, the n^+-collector is incapable of hole injection into the base.

carrier injection into a region that behaves like an insulator, so that the current is largely determined by the space charge density. An estimate of its magnitude can be made for a power transistor having a base width of 10 μm, an electron mobility of 1500 cm^2/V-sec, and a punchthrough voltage of 500 V. Substituting into (1.34) gives a collector current density of 4.5×10^5 A/cm^2, which can easily lead to destruction of the device.

1.4 DOUBLE INJECTION

It has been shown that current conduction is by single injection, provided holes injected from the anode side are unable to traverse the bar. With increasing applied voltage, however, the field across the device eventually becomes high enough to allow these holes to transit to the cathode, where they proceed to reduce the electron space charge [8]. This *double injection* process causes the space charge barrier to electron flow to be lowered, increasing the current. In addition, the hole lifetime begins to rise, permitting more holes to transit the bar. The process is thus a regenerative one, resulting in the negative resistance characteristic $A - B$, (Fig. 1.6). Let us define a threshold voltage V_T at which the hole transit time is equal to the hole lifetime under low level injection conditions. The hole transit time is approximately $\dfrac{L^2}{\mu_p V_T}$, so that

$$V_T \cong \frac{L^2}{\mu_p \tau_{p,\text{low}}} \tag{1.41}$$

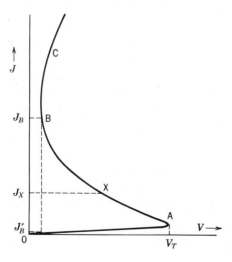

Fig. 1.6   *V-I* Characteristics for double injection.

Note that at low levels, $\tau_{n,\text{low}}$ was essentially infinite, whereas $\tau_{p,\text{low}}$ was very short. Once the threshold voltage is exceeded, however, $\tau_{p,\text{low}}$ rises with the buildup of current, whereas $\tau_{n,\text{low}}$ falls. Eventually, the high level condition is reached at point B in Fig. 1.6, where $\tau_p = \tau_n$. Two-carrier recombination-limited current flow conditions now prevail, and the insulator behaves like a low resistivity semiconductor. The negative resistance region terminates at this point.

Eventually, at sufficiently high current levels, the injection rate exceeds the recombination rate, making it possible for space charge to build up once again. It can be shown [8] that this new space charge limited region, beyond C in Fig. 1.6, can be characterized by a $J \alpha V^3$ law.

Many workers have studied single and double injection phenomena in detail [9–12] at both the theoretical and experimental levels. Furthermore, many modifications have been made to the simple theory outlined here. A discussion of these modifications is beyond the scope of this book.

1.4.1 Avalanche Injection

The mechanism of double injection requires that the anode contact be capable of supplying holes to the bar. This is certainly true for an ohmic contact. Double injection is possible, however, even if the anode contact is not capable of providing holes, by a process known as *avalanche injection* [13,14]. This situation is often encountered in the collector region of a transistor.

Consider again the circuit configuration of Fig. 1.4, with the single difference that an n^+-contact is made to the anode side. Again, we have conditions for single carrier conduction, but with the restriction that the anode is now incapable of injecting holes into the bar.

From (1.33) it is seen that the electric field intensity varies monotonically with distance, reaching a maximum value at the anode ($x = L$). As increasing voltage is impressed across the bar, this maximum electric field eventually reaches a point where moving carriers, colliding with atoms, can impart enough energy to valence band electrons to allow their promotion to the conduction band. The net result of this process is the generation of hole-electron pairs by a mechanism called *avalanche multiplication*.

Avalanche multiplication is considered in detail in Chapter 2. For the present, however, its net result is the generation of hole-electron pairs at the anode when the electric field exceeds a critical value ($\cong 1.5 \times 10^5$ V/cm for silicon). The electrons are readily collected by the n^+-anode. On the other hand, the holes are injected, as it were, into the semiconductor bar.

The *avalanche injection* process is usually initiated at threshold voltages high enough to allow double injection to occur as soon as the holes are made available, again leading to a negative resistance characteristic of the type illustrated in Fig. 1.6.

The process of avalanche injection can also be initiated in an extrinsic semiconductor bar in which current conduction normally occurs under space charge neutral conditions. Consider, for example, an n-type bar of length L, bounded by ohmic contacts. At low electric fields, the $V - J$ characteristic of such a bar is linear, so that

$$\frac{dJ}{dV} = \frac{q \mu_n N_d}{L} \qquad (1.42)$$

where N_d is the doping concentration and μ_n is the electron mobility. Space charge neutral conditions exist during this mode of operation.

As the current is increased, the value of electric field becomes so large that the mobility is no longer constant. Eventually, at an electric field of about 10^4 V/cm in silicon, the electron drift velocity saturates at a limiting value $v_{\lim} = 10^7$ cm/sec. This corresponds to a critical current density J_0 where

$$J_0 = q v_{\lim} N_d \qquad (1.43)$$

For current densities above J_0, a negative excess charge is stored throughout the n-region, such that

$$\rho = q N_d - \frac{J}{v_{\lim}} \qquad (1.44)$$

and $J > J_0$. Thus operation in the saturated velocity region results in space charge limited flow for the excess current. With increasing current, this eventually leads to avalanche injection as described here.

1.4.2   Impact Ionization

A negative resistance characteristic can also be initiated at any point within a semiconductor bar by a process termed *impact ionization*. Consider, as before, an n-type semiconductor bar in which conduction is by single carrier injection. A constant voltage V_A is applied across its contacts (Fig. 1.7a). Assume that at some point X in the bar, there exists a nonuniformity, created perhaps by a metal or an oxide precipitate during processing. Let us further assume that the resistivity of the bar is sufficiently high that the bar can support a significant electric field. The principal effect of such a nonuniformity is to create a highly localized concentration of the electric field in its vicinity, because of distortion of the potential lines.

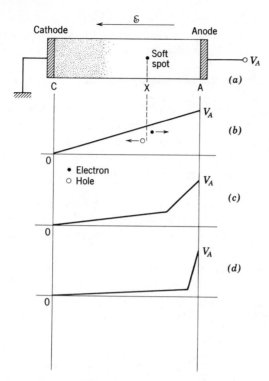

Fig. 1.7   Impact ionization at a "soft spot."

Initially currrent flow is by single injection, and it creates an electron space charge region at the cathode. As the applied voltage is raised, local breakdown by impact ionization [15] eventually occurs at this "soft spot," giving rise to hole-electron pair generation at X.

The voltage across the bar before impact ionization is shown in Fig. 1.7b. Upon impact ionization, holes move toward the cathode and reduce the electron space charge in this region. Thus double injection conditions prevail, and the resistivity of the bar between C and X falls with increasing current. Consequently the electric field across the X–A region increases, resulting in a shift of the ionization process toward the anode (Fig. 1.7c). The eventual voltage distribution approaches that of Fig. 1.7d. At this point most of the bar is flooded with electrons and holes, and two-carrier recombination limited current flow conditions prevail.

The $V-J$ characteristic of a bar in the presence of impact ionization thus is very similar to that obtained with double injection. This is to be expected, since impact ionization leads to double injection and the characteristic negative resistance features shown in Fig. 1.6.

We now consider some of the consequences of double injection, avalanche injection, and impact ionization, particularly as they affect the behavior of high power semiconductors.

1.4.3 Current Filaments and Microplasmas

All the preceding processes lead to the negative resistance type of behavior represented in Fig. 1.6. A device exhibiting such behavior is said to be *open circuit stable*, and is often referred to as having a CCNR (current-controlled negative resistance) characteristic. It can be shown [16] that this will lead to current flow by means of filamentary conduction between the anode and cathode.

Consider a bar of cross section A, with the $V - J$ characteristic shown in Fig. 1.6. The bar is biased at a point X in the negative resistance region, corresponding to a current density of J_X; that is, the current through the bar is AJ_X. The terminal voltage eventually reaches the minimum value V_B, where the negative incremental resistance becomes zero, corresponding to stable operation. This is accomplished by current flow at a density J_B through a cross-sectional area a, and at a density J_B' through the remainder. Thus

$$J_X A = J_B a + J_B' (A - a) \tag{1.45}$$

or

$$a = \frac{A(J_X - J_B')}{J_B - J_B'} \tag{1.46}$$

Since J_B is much larger than J_X or J_B', it follows that most of the current flows through a small cross section a. Thus the semiconductor becomes electrically heterogeneous as excess current flows through a small cross section. For this condition, the resistivity of the core is lower than that of the surrounding material, tending to shunt more of the total current through it, further lowering its resistivity, and further increasing that of the surrounding material. This process continues until almost all the current flows through this core in the form of a current filament. This is a general property of all CCNR devices and does not depend on the specific way in which this characteristic arises.

The existence of CCNR characteristics and of filamentary conduction has been confirmed by a number of workers [9,17], for a variety of semiconductor materials. Such filaments have often been termed *microplasmas* since they consist of quasi-neutral regions, approximately 1–10 μm in diameter [18], flooded with holes and electrons in almost equal number. Their light emission characteristics have been determined [19] and in general show a broad peak around the energy gap, together with a black body radiation characteristic caused by the localized increase of temperature in this vicinity.

In steady state, the rate of hole-electron pair production in a microplasma is equal to the rate of loss of carriers from the filament by diffusion. Thus an increase in the rate of pair production (produced, e.g., by raising the applied voltage), will be balanced by an increase in filament diameter, until a new steady state condition is reached. Consequently the microplasma is inherently stable. This is true as long as circuit conditions can keep local temperatures below the point at which additional mechanisms for hole-electron pair production are not present. These effects are dicusssed in the following sections.

1.5   THERMAL INSTABILITY

We have noted that a microplasma can be relatively stable, since any increase in its current density can result only from an increase in the electric field that supports it. This is true for the processes described here. At elevated temperatures, however, thermal generation of hole-electron pairs can significantly increase the current flow. This is a regenerative process, since increased current flow is accompanied by an increase in filament temperature [20]. Eventually thermal runaway occurs, with destructive failure of the device. The temperature at which this process is initiated is of interest in determining the safe operating limits of a power device.

1.5.1   The Intrinsic Temperature

In silicon, the generation of hole-electron pairs by thermal means results in the formation of n_i electrons and holes per cubic centimeter, where

$$n_i = 1.69 \times 10^{19} \left(\frac{T}{300} \right)^{3/2} e^{-0.55/kT} \qquad (1.47)$$

Typically the intrinsic concentration is $1.38 \times 10^{10}/\text{cm}^3$ at $300°\text{K}$, and it is negligible compared to the doping level. However n_i increases rapidly with temperature, doubling every $11°\text{C}$ approximately. At high temperatures, therefore, thermal generation can be the dominant process of hole-electron pair production.

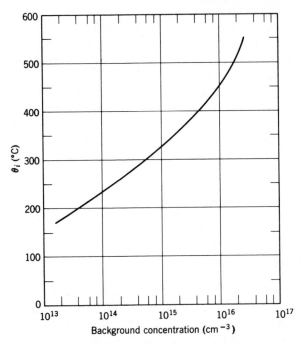

Fig. 1.8 Intrinsic temperature versus doping level.

The carrier concentration due to thermal effects becomes equal to the background concentration at the *intrinsic temperature* θ_i, which is presented in Fig. 1.8 as a function of background concentration. Below θ_i, the carrier concentration in silicon is relatively temperature independent. Above θ_i, however, it rises exponentially with temperature. Thus the consequence of localized heating in a current filament depends on whether the filament temperature exceeds θ_i. The filament is inherently stable below the intrinsic temperature and is maintained by the electric field. Beyond θ_i, however, the generation of hole-electron pairs can advance rapidly and independently of the electric field, accompanied by a rapid local rise in temeprature. Typically this is accompanied by a rise in current density by as much as 2 to 3 orders of magnitude, resulting in the formation of a giant or "*meso*" *plasma*.

An additional factor that aids mesoplasma formation is the behavior of the thermal conductivity of silicon, which varies inversely with absolute temperature. Thus the buildup of temperature during mesoplasma formation is assisted by the simultaneously falling thermal conductivity of the silicon.

An important generalization can now be made: *any* inhomogeneity that results in a localized temperature in excess of the intrinsic value can lead to mesoplasma formation and thermal runaway. The inhomogeneity can come about as a material property, because of structural or doping imperfections; alternately, an ideally uniform semiconductor can become electrically inhomogeneous by the formation of current filaments, if it has a CCNR characteristic.

In principle, a mesoplasma can be stabilized by appropriately controlling the power delivered to it. In practice, however, such control is extremely difficult to exert, since core temperatures are very sensitive to power input under mesoplasma conditions.

1.5.2 Mesoplasmas and Second Breakdown

A typical mesoplasma takes the form [21, 22] of a strongly glowing red spot in the central regions of a device, having an average temperature of 650°C and a peak core temperature in excess of 1000°C. Spot diameters of 25–75 μm have been measured in a number of devices. However there is considerable uncertainty about these figures because of the difficulty of making measurements of this type.

The establishment of a mesoplasma usually results in irreversible damage to the device. This failure mode is often referred to as *second breakdown*, in distinction to avalanche or *first breakdown*, which is often accompanied by microplasma formation and is reversible.

The damage associated with second breakdown can come about in a number of ways, as follows:

1. The silicon melts if the core temperature reaches 1412°C. Melting often takes the form of circular holes, many microns in diameter.
2. The device is destroyed when the surface temperature exceeds the eutectic point associated with the silicon and the surface metal with which it is in contact. This melting is accompanied by the transport of contact metal into the mesoplasma region and a fall in the melting temperature of the core. Studies of second breakdown have shown that this is a common occurrence. In fact, the onset of second breakdown can be delayed by the use of refractory metal contact such as molybdenum since the molybdenum-silicon system has a high eutectic temperature (1410°C).
3. Device failure occurs because of crystal damage and cracking, caused by thermal shock during the rapid buildup of local temperature.

Studies of breakdown in high voltage junctions [23] reveal that the mesoplasma core is usually at or near the metallurgical junction. Damage in the form of a treelike structure begins at this core and extends with each thermal overload until the entire depletion layer is shorted out. Device failure thus takes the form of a successive lowering of the breakdown voltage. This overall process is field initiated but thermally terminated. This is true, in general, for all devices that fail in second breakdown.

1.5.2.1 Steady State Characteristics [24]

A steady state analysis of current flow in a mesoplasma can be made by considering an infinitely large area, uniform disk of semiconductor material across which a voltage is applied. Assume that current has been constricted to flow through a region of minimum radius r_0, which is much smaller than the physical dimensions of the device (Fig. 1.9a). Assuming symmetry, let S_m be the surface of maximum temperature θ_m, as well as the surface of reference potential. Isothermal surfaces S and S' are also shown, for which the temperature and potentials are (θ, ϕ) and $(\theta - d\theta, \phi + d\phi)$, respectively (Fig. 1.9b).

If J is the current density at any point, and ρ is the electrical resistivity,

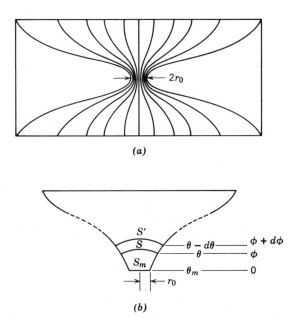

(a)

(b)

Fig. 1.9 Current constriction during mesoplasma formation. Copyright © 1966 by the Institute of Electrical and Electronic Engineers, Inc. Reprinted with permission from Khurana et al. [24].

then

$$J = -\frac{1}{\rho} \operatorname{grad} \phi \tag{1.48}$$

The flow of heat per unit area is

$$Q = \phi \cdot J - K \operatorname{grad} \theta \tag{1.49}$$

where K is the thermal conductivity and θ is the temperature. But $\operatorname{div} J = \operatorname{div} Q = 0$, so that

$$\operatorname{div}(-\phi \operatorname{grad} \phi - \rho K \operatorname{grad} \theta) = 0 \tag{1.50}$$

This equation is satisfied if

$$\phi \operatorname{grad} \phi + \rho K \operatorname{grad} \theta = 0 \tag{1.51}$$

Integrating from S_m to S,

$$\int_0^\phi \phi \operatorname{grad} \phi = -\int_{\theta_m}^\theta \rho K \operatorname{grad} \theta \tag{1.52}$$

so that

$$\phi^2 = 2 \int_\theta^{\theta_m} \rho K \, d\theta \tag{1.53}$$

Experimental values of the ρK product for silicon appear in Fig. 1.10.

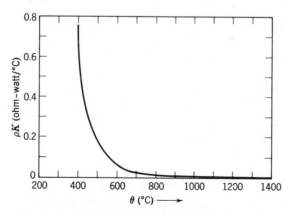

Fig. 1.10 Variation of ρK with temperature, for silicon. Copyright © 1966 by the Institute of Electrical and Electronic Engineers, Inc. Reprinted with permission, from Khurana et al. [24].

Note that for θ below 500°C, this product is a steeply falling function of temperature.

For all practical purposes, the mesoplasma is bounded on either side of the core region by isothermals at θ_i. Therefore integration of (1.53) must be carried out to surfaces that are at this temperature. The region beyond this surface can be ignored and creates no potential drop, since it was assumed to have an infinitely large diameter. For finite diameter devices, however, this adds an additional resistive drop, which is determined by doping level. Note that θ is the same for either side of the core in this analysis, since uniformly doped material is assumed. On the other hand, the analysis can be extended to mesoplasmas in nonuniformly doped structures such as $p-n$ junctions, if appropriate values of intrinsic temperature are chosen for the p- and n-semiconductor regions. In a diode, the mesoplasma core is located at the point of maximum electric field, which is usually close to the metallurgical junction.

Figure 1.11 shows the solution of (1.53). It is seen that the magnitude of ϕ is almost independent of θ_m for $\theta_m \gg \theta_i$. This is usually true for power devices that have good heat sinks. For such devices, the surface temperature is generally below θ_i; thus the actual value of the potential drop across the mesoplasma region is relatively independent of device dimensions $(2\phi \cong 20 \mathrm{V})$.

The foregoing analysis has been extended to provide an estimate of the core radius for a mesoplasma current of I_m. The results of this analysis are given in Fig. 1.12. This result has been crudely verified in practice [22].

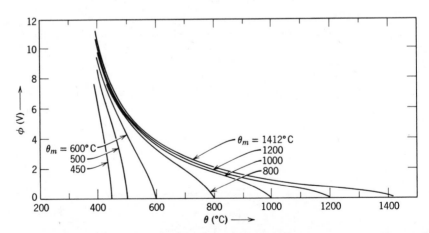

Fig. 1.11 Solution of (1.53) for different values of peak temperature. Copyright ©
1966 by the Institute of Electrical and Electronic Engineers, Inc. Reprinted with
permission from Khurana et al. [24].

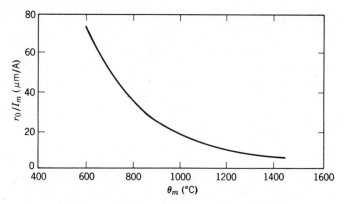

Fig. 1.12 Core radius as a function of peak temperature.

An estimate can now be made of the $V - I$ characteristic of a device in its mesoplasma regime. Although an increase of current through the device results in an increase of θ_m, the voltage across a mesoplasma is relatively constant as long as the surface temperature of the device is much less than θ_i. As a result, a device with a perfect heat sink exhibits a constant voltage drop ($2\phi \cong 20\,V$, as seen from Fig. 1.11) independent of current. If the heat sink is not perfect, the surface temperature rises and the terminal voltage falls, tending to keep the power dissipated to a constant value in the limiting case. In a practical device, therefore, the $V - I$ characteristic falls somewhere between the constant voltage and the constant power curves (Fig. 1.13).

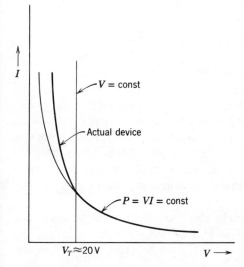

Fig. 1.13 Static characteristic of a device in the mesoplasma mode.

1.5.2.2 Initiation Time [25]

The initiation time of a mesoplasma can now be calculated if a number of simplifying assumptions are made. We begin by assuming a spherical mesoplasma core, of radius r_0, imbedded in an infinite semiconductor bulk that is maintained at an ambient temperature of T_A. A constant power P is suddenly applied to this core for a time interval τ_D, at which point irreversible damage occurs. This is characterized by a core temperature T_M, at or near the melting point of the material (1412°C for silicon).

Finally, it is assumed that the process can be characterized by average values of specific heat and thermal conductivity over the temperature range from 25 to 1412°C. Thus let C be the average specific heat [J/g-°C], and K the average thermal conductivtiy, in W/cm-°C. For silicon, $C \cong 0.935$ and $K \cong 0.415$ over the temperature range from 25 to 1412°C.

Solutions to the heat flow equation, based on these assumptions, are available in the literature [26]. In approximate form,

$$T_M - T_A \cong \frac{r_0^2 P}{2K\left[(4/3)\pi r_0^3\right]} \frac{4K\tau_D/r_0^2\rho C}{2 + 3\left(4K\tau_D/r_0^2\rho C\right)^{1/2} + 4K\tau_D/r_0^2\rho C} \quad (1.54)$$

where ρ is the density (2.33 g/cm^3 for silicon). Rearranging terms,

$$P \cong (T_M - T_A)\left[\frac{CV}{\tau_D} + S\left(\frac{\rho C K}{\tau_D}\right)^{1/2} + \left(\frac{8}{3}\right)\pi r_0 K\right] \quad (1.55)$$

where S and V are the surface area and volume of the mesoplasma core, respectively.

Equation 1.55 indicates two important features. When the initiation time for the mesoplasma is short compared to the thermal relaxation time, $P\tau_D =$ constant. Here the mesoplasma core is essentially out of communication with the bulk material. This situation arises in extremely small devices or in devices that are subject to large overload conditions. For longer time delays, where diffusion of heat from the core into the bulk is significant, $P\sqrt{\tau_D} =$ constant. Eventually the steady state condition is reached, where the temperature of the entire material varies uniformly with the application of power.

The present theory, although simple, has been experimentally verified by a number of workers. Thus $P\tau_D =$ constant behavior has been observed for devices with extremely short delay time [27], on the order of 0.01–1.0 μsec.

On the other hand, measurements on devices with longer delay times [28], on the order of 20–500 μsec, have shown that the $P\sqrt{\tau_D}$ = constant law holds for these situations.

There is considerable lack of uniformity [29,30] with respect to the actual temperature at which second breakdown is thought to occur. Some workers define its onset at the point at which the core temperature reaches θ_i. Yet others consider temperatures at or near the melting point of silicon to be more realistic. The concept of an intrinsic temperature has also been questioned, and some evidence [31] indicates that the mesoplasma initiation temperature is not strictly related to the doping level but is also a function of the input pulse power. Finally, it is not certain whether a mesoplasma can take the form of a stable sphere; alternate models have been proposed [32] in which its core becomes an ellipsoid whose major axis expands until it completely bridges the device.

A number of initiating mechanisms that can bring about the onset of a mesoplasma in semiconductor devices are related to the physics of device operation. These are discussed in subsequent chapters. In all cases, however, the mesoplasma represents a terminal phase that leads to irreversible damage to the device.

1.6 REFERENCES

1. W. Shockley and W. T. Read, "Statistics of Recombination of Holes and Electrons," *Phys. Rev.*, **87**, pp. 835–842 (1952).

2. R. N. Hall, "Electron-Hole Recombination in Silicon," *Phys. Rev.*, **87**, pp. 387–392 (1952).

3. J. L. Moll, *Physics of Semiconductors*, McGraw-Hill Book Co., New York, 1964.

4. J. Cornu et. al., "Analysis and Measurement of Carrier Lifetimes in the Various Operating Modes of Power Devices," *Solid State Electron.* **17**, No. 10, pp. 1099–1106 (1974).

5. K. G. Svantesson et al., "Recombination in Strongly Excited Silicon," *Solid State Commun.*, **9**, No. 3, pp. 213–216 (1971).

6. J. D. Beck and R. Conradt, "Auger-Rekombination in Si," *Solid State Commun.*, **13**, No. 1, pp. 93–95 (1973).

7. M. A. Lampert, "Injection Currents in Insulators," *Proc. IRE*, **50**, No. 8, pp. 1781–1795 (1962).

8. M. A. Lampert, "Double Injection in Insulators," *Phys. Rev.*, **125**, pp. 126–141 (1962).

9. N. Holonyak, "Double Injection Diodes and Related DI Phenomena in Semiconductors," *Proc. IRE*, **50**, No. 12, pp. 2421–2428 (1962).

10. K. L. Ashley and A. G. Milnes, "Double Injection in Deep-Lying Impurity Semiconductors," *J. Appl. Phys.*, **33**, No. 2, pp. 369–374 (1964).

11. J. L. Wagener and A. G. Milnes, "Double-Injection Experiments in Semi-Insulating Silicon Diodes," *Solid State Electron.*, **8**, pp. 495–507 (1965).

12. A. P. Ferro and S. K. Ghandhi, "Properties of Gallium Arsenide Double Injection Devices," *J. Appl. Phy.*, **42**, No. 10, pp. 4015–4024 (1971).

13. P. L. Hower and V. G. K. Reddi, "Avalanche Injection and Second Breakdown in Transistors," *IEEE Trans. Electron Devices*, **ED-17**, No. 4, pp. 320–335 (1970).

14. G. Vitale, "Negative Resistance in Si Bulk Devices," *Solid State Electron.* **18**, No. 12, pp. 1123–1130 (1975).

15. J. N. Park et al., "Avalanche Breakdown Effects in Near-Intrinsic Silicon and Germanium," *J. Appl. Phys.*, **38**, No. 13, pp. 5343–5351 (1967).

16. B. K. Ridley, "Specific Negative Resistance in Solids," *Proc. Phys. Soc. (London)*,**82**, p. 954 (1963).

17. A P. Ferro and S. K. Ghandhi, "Observations of Current Filaments in Chromium-Doped GaAs," *Appl. Phys. Lett.*, **16**, No. 5, pp. 196–198 (1970).

18. M. W. Muller and H. Guckel, "Negative Resistance and Filamentary Currents in Avalanching Silicon p^+-i-n^+ Junctions," *IEEE Trans. Electron Devices*, **ED-15**, No. 8, pp. 560–568 (1968).

19. W. M. Portnoy and F. R. Gamble, "Fine Structure and Electromagnetic Radiation in Second Breakdown," *IEEE Trans Electron Devices*, **ED-11**, No. 10, pp. 470–478 (1964).

20. T. Agatsuma,"Turn Over Phenomena in n-v-n Si Devices and Second Breakdown in Transistors," *IEEE Trans. Electron Devices*, **ED-13**, No. 11, pp. 748–753 (1966).

21. A. C. English, "Mesoplasmas and 'Second Breakdown' in Silicon Junctions," *Solid State Electron.*, **6**, pp. 511–521 (1963).

22. A. C. English, "Physical Investigation of the Mesoplasma in Silicon," *IEEE Trans. Electron Devices*, **ED-13**, No. 8/9, pp. 662–667 (1966).

23. I. Thomson and E. L. G. Wilkinson, "Destructive Breakdown in Large Area Phosphorus-Diffused High Voltage n^+-p Junctions," *Solid State Electron.*, **10**, pp. 983–989 (1967).

24. B. S. Khurana et al., "Thermal Breakdown in Silicon $p-n$ Junction Devices," *IEEE Trans. Electron Devices*, **ED-13**, No. 11, pp. 763–770 (1966).

25. D. M. Tasca, "Pulse Power Failure Modes in Semiconductors," *IEEE Trans Nucl. Sci.*, **NS-17**, pp. 364–372 (1970).

26. H. S. Carslaw and J. C. Jaeger, *Conduction of Heat in Solids*, Oxford University Press, London, 1959.

27. D. K. Ferry and A. A. Dougal, "Input Power Induced Thermal Effects Related to Transition Time Between Avalanche and Second Breakdown in $p-n$ Silicon Junctions," *IEEE Trans. Electron Devices*, **ED-13**, No. 8/9, pp. 627–629 (1966).

28. H. Melchior and M. J. O. Strutt, "Secondary Breakdown in Transistors," *Proc. IEEE*, **52**, No. 4, pp. 439–440 (1964).

29. W. B. Smith et al., "Second Breakdown and Damage in Junction Devices," *IEEE Trans. Electron Devices*, **ED-20**, No. 8, pp. 731–744 (1973).

30. D. H. Pontius et al., "Filamentation in Silicon-on-Sapphire Homogeneous Thin Films," *J. Appl. Phys.*, **44**, pp. 331–340 (1973).

31. R. A. Sunshine and M. A. Lampert, "Second Breakdown Phenomena in Avalanching Silicon-on-Sapphire Diodes," *IEEE Trans. Electron Devices*, **ED-19**, No. 7, pp. 873–885 (1972).

32. F. Weitzsch, "A Discussion of Some Known Physical Models for Second Breakdown," *IEEE Trans. Electron Devices*, **ED-13**, No. 11, pp. 731–734 (1966).

1.7 PROBLEMS

1. A bar of n-type silicon is illuminated with F photons/sec of penetrating light. Develop an expression for the growth of excess electrons and holes in the bar, assuming that each photon creates a hole-electron pair. Assume low illumination levels.

2. In n-type silicon, gold can be represented by a single acceptor, located at 0.57 eV above the valence band. The capture rate constants for this deep impurity are $\alpha_n = 1.65 \times 10^{-9}$ cm^3/sec and $\alpha_p = 1.15 \times 10^{-7}$ cm^3/sec.

   A $p^+ - n$ diode, with a concentration of 10^{14} atoms/cm^3 in the n-region, is doped uniformly with 3×10^{13} atoms/cm^3 of gold. Compute the values of $\tau_{p,\text{low}}$, $\tau_{p,\text{high}}$, and τ_{sc} for this diode, at 27°C.

3. Repeat Problem 2, for 125°C.

4. A region of a semiconductor is flooded with injected carriers at an injection level of 5×10^{17}/cm^3. The high level lifetime for this region, in the absence of Auger recombination, is 10 μsec. Compute the true lifetime.

5. Repeat Problem 4, assuming a high level lifetime of 100 μsec.

6. A piece of n-type silicon, doped to 10^{14}/cm^3, has a hole diffusion constant of 12 cm^2/sec. Sketch the ambipolar diffusion constant for injected carrier concentrations from zero to 10^{17}/cm^3.

7. Repeat Problem 6 for p-type silicon.

8. Explain physically why a punched-through p^+-ν-n^+ diode cannot be described by single carrier injection.

CHAPTER

2

The Reverse-
Biased Diode

CONTENTS

A POWER SEMICONDUCTOR device must be capable of sustaining a high voltage when operated in its nonconducting or OFF state. Ideally, this should be accomplished with a low reverse leakage current, to minimize the dissipated power. In all devices, this voltage is supported by the depletion layer associated with a reverse-biased p-n junction. Reverse voltages range from as low as 5 V for a typical emitter-base junction of a high speed transistor, to many kilovolts for power rectifier diodes and thyristors.

Alloying is often used for the gate-cathode junction of thyristors, where an antimony-doped, gold-germanium eutectic preform is used for the n-region (cathode), with a junction depth of a few microns. Characteristic of these junctions is their inability to support high reverse voltages. Typically they are used in a forward-biased mode and are closely approximated in characteristics by an abrupt doping profile.

The typical low voltage diode is made by conventional diffusion techniques, and is usually 2–5 μm deep. Phosphorus and boron are commonly used as the n- and p-type dopants, respectively. High voltage junctions are also made by diffusion, with depths ranging from 10 to more than 200 μm. Because of these large diffusion depths, the junctions cannot be considered to be abrupt. Typical doping profiles are complementary error function and Gaussian; linearly graded junction theory can often be used to explain their characteristics. Gallium and phosphorus are the commonly used p- and n-type dopants for these junctions.

This chapter considers the different types of abrupt and graded junctions used in high power semiconductors and their behavior under reverse bias conditions. The properties of diffused junctions are also considered, especially as derivable from the linear junction model.

The uniform introduction of dopant into a slice of semiconductor material results in a parallel plane junction. The use of photo masking processes has gained increasing importance for the delineation of diffused junctions in power semiconductors, particularly in situations of many devices being made on each silicon slice. Here, the parallel plane junction takes on a cylindrical shape at its edges, with a radius slightly less than the junction depth (approximately 80%). As a consequence, the boundaries of the depletion layer also become curved in this region. The effect of depletion layer curvature on the breakdown voltage of these junctions is also considered. In addition, techniques are described for controlling this curvature to obtain improved voltage breakdown characteristics in high voltage junctions.

2.1 REVERSE CURRENT

Reverse leakage current may be caused by surface contamination, by the diffusion of minority carriers from the neutral regions to the depletion layer edge, or by charge generation at deep levels within the depletion layer. Surface leakage is minimized by the use of clean processing techniques, by junction coating with glasses or silicone resins, and by designing the device to ensure that surface fields are kept below internal fields (see Sec. 2.6). Diffusion-limited leakage currents can be shown to be negligible for silicon diodes operating below 100°C. Consequently, the primary cause of leakage current (in this temperature range) is charge generation at deep levels in the depletion layer.

Figure 2.1 shows the energy band configuration for a p-n junction with a reverse voltage of magnitude V. For this condition, consider the behavior at a recombination center within the high field depletion region (the

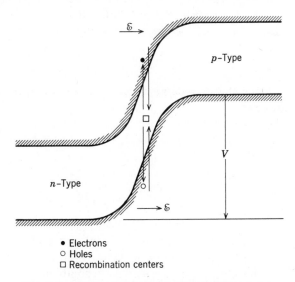

• Electrons
○ Holes
□ Recombination centers

Fig. 2.1 Charge generation in the depletion layer.

direction of this field is indicated in the figure). Assume that the recombination centers are approximately midway between the valence and conduction bands. Band-to-band generation effects can be neglected, compared to generation at these centers.

The generation rate for electrons (or holes) is directly proportional to the number of appropriately charged centers at the recombination level and to the number of available states in the conduction (or valence) bands. Thus as far as these rates are concerned, the situation inside the depletion layer is not very different from that outside it.

The capture rate, on the other hand, is proportional to the number of mobile carriers in the valence and conduction bands, and to the number of recombination centers of the appropriate charge state. Since the number of mobile carriers in the depletion layer is very small (indeed, it is ordinarily assumed that the region has *no* mobile carriers), capture rates are considerably lower than those encountered outside the depletion layer.

Thus the presence of deep-lying levels in the depletion layer gives rise to the net *generation* of charge, with each center emitting electrons and holes in succession. Upon generation, these charges are swept out of the depletion layer under the influence of the high electric field associated with this region. For every hole-electron pair generated in this manner, only one mobile carrier is transported past any given reference plane and delivered to the external circuit.

The magnitude of the leakage current may now be determined for steady state conditions. Let U be the generation rate for carriers within the depletion layer. Then the leakage current is qU amperes per unit volume of the depletion layer. Writing W as the width of this depletion layer gives

$$J_{sc} = qUW \tag{2.1}$$

where J_{sc} is the space charge generated leakage current density in amperes per square centimeter.

From the Sah-Noyce-Shockley theory of recombination [1], modified to include the fact that both n and p are negligible within the depletion layer, we obtain [see (1.18)]

$$U = \frac{n_i^2}{n_1\tau_{p0} + p_1\tau_{n0}} \equiv \frac{n_i}{2\tau_{sc}} \tag{2.2}$$

where n_i is the intrinsic concentration, τ_{sc} is the space charge generation lifetime, and p_1, n_1 are the equilibrium values of the hole and electron concentrations if the Fermi level were at the recombination level. Assuming mid-gap centers

$$U \cong \frac{n_i}{2\tau_{sc}} \tag{2.3}$$

so that

$$J_{sc} \cong \frac{qn_i W}{2\tau_{sc}} \tag{2.4}$$

From (2.4) it is seen that the reverse current density varies directly as the volume of the depletion layer and inversely as the space charge generation lifetime. This suggests an important reason for maintaining long lifetime during device processing of high voltage junctions, where wide depletion layers are common.

For mid-gap recombination centers, the charge generation current J_{sc} is directly proportional to n_i; and thus, it approximately doubles with every 11°C rise in temperature. Figure 2.2 plots the charge generation current density for this type of center.

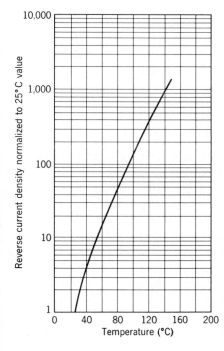

Fig. 2.2 Leakage current density versus temperature. From *Transistor Engineering* by A. Phillips. Copyright 1962. Used with permission of McGraw-Hill Book Company.

The diffusion-limited reverse current density becomes significant at junction temperatures above 100°C. Such temperatures are often exceeded in power semiconductors. Consider a reverse-biased diode, having p- and n-regions of widths W_P and W_N, measured from the depletion layer edges. The minority carrier concentrations at these edges are zero. Let $\bar{n}_P$ and $\bar{p}_N$ be equilibrium concentrations of minority carriers in the p- and n-regions, respectively. Finally, assume that these regions are "short" compared to the appropriate minority carrier diffusion lengths. For diffusion-limited current conditions, the reverse current densities are then given by

$$J_n = qD_n \frac{dn}{dx} = \frac{qD_n\bar{n}_P}{W_P} \tag{2.5a}$$

and

$$J_p = -qD_p \frac{dp}{dx} = \frac{qD_p\bar{p}_N}{W_N} \tag{2.5b}$$

where D_n and D_p are electron and hole diffusion constants.

The diffusion-limited leakage current density is given by the sum of these terms. For either region, it can be shown that the base width must be replaced by the minority carrier diffusion length, if "long" regions are considered.

2.2 AVALANCHE BREAKDOWN

Consider a semiconductor subjected to an increasing electric field. Eventually, a point is reached when mobile carriers attain their terminal drift velocity (10^7 cm/sec for electrons and 6.5×10^6 cm/sec for holes in silicon). With further increase of electric field, the velocity of individual carriers exceeds their thermal velocity; that is they become "hot" carriers. At high fields, these hot carriers collide with atoms and impart enough energy to valence band electrons to promote them to the conduction band, resulting in hole-electron pair generation. This process of *impact ionization* is multiplicative. Each newly generated hole or electron also is involved in the ionization of further hole-electron pairs. Avalanche breakdown occurs when this process attains an infinite rate.

The ionization process can be characterized [2] by a pair of ionization coefficeints, α_p and α_n, for holes and electrons, respectively. Thus

α_p = number of hole-electron pairs produced per hole per

centimeter traveled in the direction of the $\mathscr{E}$ field

α_n = number of hole-electron pairs produced per electron per

centimeter traveled in the direction of the $\mathscr{E}$ field

For silicon,* at $\mathscr{E}$ fields between 5×10^4 and 8×10^5 V/cm,

$$\alpha_n = ae^{-b/\mathscr{E}} \tag{2.6a}$$

$$\alpha_p = \gamma ae^{-b/\mathscr{E}} \tag{2.6b}$$

where

$$a = 1.6 \times 10^6 /\text{cm} \tag{2.7}$$

$$b = 1.65 \times 10^6 \text{ V}/\text{cm} \tag{2.8}$$

$$\gamma = 0.344 \tag{2.9}$$

Consider a single hole-electron pair that has been generated at a point x, within the depletion layer of a p-n diode [4]. The hole will be swept by the electric field toward the p-region and will encounter an average of $\alpha_p dx$ ionizing collisions when traversing a distance dx. In like manner, the electron will be swept toward the n-region, with an average of $\alpha_n dx$ ionizing collisions over the interval dx. Secondary pairs produced in this manner will generate chains of collisions as they traverse the depletion layer. If $M(x)$ is the average total number of pairs generated in the depletion layer as a result of a single initial pair generated at x, then

$$M(x) = 1 + \int_0^x \alpha_n M(x')dx' + \int_x^W \alpha_p M(x')dx' \tag{2.10}$$

where W is the width of the depletion layer. Differentiating,

$$\frac{dM(x)}{dx} = (\alpha_n - \alpha_p) M(x) \tag{2.11}$$

*More extensive data on ionization coefficients are given in Reference 3.

Solving, and setting

$$M(x) = M(0) \quad \text{for} \quad x = 0 \tag{2.12}$$

it is seen that

$$M(x) = M(0) \exp\left[\int_0^x (\alpha_n - \alpha_p) dx' \right] \tag{2.13}$$

Substituting into (2.10) for $x = 0$,

$$1/M(0) = 1 - \int_0^W \alpha_p \exp\left[\int_0^x (\alpha_n - \alpha_p) dx' \right] dx \tag{2.14}$$

so that

$$M(x) = \frac{\exp\left[\int_0^x (\alpha_n - \alpha_p) dx' \right]}{1 - \int_0^W \alpha_p \exp\left[\int_0^x (\alpha_n - \alpha_p) dx' \right] dx} \tag{2.15}$$

Breakdown occurs when $M(x)$ become infinite, that is, when

$$\int_0^{W'} \alpha_p \exp\left[\int_0^x (\alpha_n - \alpha_p) dx' \right] dx = 1 \tag{2.16}$$

Here W' is the width of the depletion layer at breakdown.

This expression applies to hole-electron pair multiplication in the depletion layer, *regardless* of whether it results from carriers that enter the depletion layer by diffusion through the neutral regions or from carriers produced by charge generation at deep levels within the depletion layer.

2.3 MULTIPLICATION FACTOR

The total reverse current of a *p-n* diode is made up of the combined effect of multiplication of carriers coming from the quasi-neutral end regions, and of carriers generated at deep levels in the depletion layer. It has been shown that the condition for avalanche breakdown is the same for all these; that is, each component becomes infinite at the same breakdown voltage. However detailed solutions of (2.15) for the individual components demonstrate that each term approaches infinity at a different rate. The results of these calculations are given below.

Let J_{n0} and J_{p0} be electron and hole current densities of carriers entering a depletion layer from the quasi-neutral p- and n-regions, respectively. Furthermore, let J_{sc} be the space charge generation current density in this depletion layer. Define multiplication factors M_n, M_p, and M_{sc} corresponding to these terms. Then it can be shown [5] that, for high voltage diodes

$$M_n \cong \frac{1}{1-(V_j/BV)^4} \qquad (2.17)$$

and

$$M_p \cong \frac{1}{1-(V_j/BV)^6} \qquad (2.18)$$

where V_j is the reverse voltage supported by the junction, and BV is the breakdown voltage. In addition, *and this is extremely significant*, solution of (2.15) also shows that $M_{sc} \cong M_n$ for an n^+-p diode, but $M_{sc} \cong M_p$ for a p^+-n diode. Thus although the charge generation current is the dominant leakage term in silicon diodes, its variation with voltage is dictated by the behavior of the diffusion-limited current. These results indicate that the reverse current of a p^+-n diode approaches infinity as a steeper function of voltage than that for an n^+-p diode. This is because the diffusion-limited current in a p^+-n diode is primarily due to holes from the n-region.

The multiplication factor can be applied to the collection efficiency of a transistor also. In an n-p-n transistor, electrons are the injected minority carriers that enter the collector-base depletion layer. The multiplication factor appropriate to this situation is M_n. In like manner, M_p is the multiplication factor for collected holes in a p-n-p transistor. Note that unlike the diode, space charge generation in the collector-base depletion layer of a transistor is negligible compared to the effects of injected minority carriers entering this region.

2.4 BREAKDOWN VOLTAGE

In Section 2.2 it was shown that breakdown occurs when

$$\int_0^{W'} \alpha_p \exp\left[\int_0^x (\alpha_n - \alpha_p)dx'\right] dx = 1 \qquad (2.16)$$

This equation can be simplified if an average ionization coefficient α_i is used, instead of two separate coefficients. Thus if the approximation* is made that

$$\alpha_n \cong \alpha_p \cong \alpha_i \cong 1.07 \times 10^6 \exp(-1.65 \times 10^6 / \mathscr{E}) \qquad (2.19)$$

(2.16) reduces to the simpler form

$$\int_0^{W'} \alpha_i \, dx = 1 \qquad (2.20)$$

The condition for avalanche breakdown can also be developed for unequal ionization coefficients. Thus noting that

$$\int_0^{W'} (\alpha_n - \alpha_p) \exp\left[\int_0^x (\alpha_n - \alpha_p) \, dx' \right] dx = \exp\left[\int_0^{W'} (\alpha_n - \alpha_p) \, dx \right] - 1 \quad (2.21)$$

and $\alpha_p = \gamma \alpha_n$, it can be shown that (2.16) reduces to

$$\int_0^{W'} \alpha_n \, dx = \frac{\ln \gamma}{\gamma - 1} \qquad (2.22)$$

or to

$$\int_0^{W'} \alpha_p \, dx = \frac{\gamma \ln \gamma}{\gamma - 1} \qquad (2.23)$$

Experimental data [6] on both abrupt and diffused diodes show that (2.20) provides a better fit than (2.22) or (2.23). Consequently the simpler form of the ionization integral will be used for calculating the breakdown voltage of p-n junctions.

2.4.1 The Abrupt Junction

An abrupt junction is obtained in alloyed and shallow diffused structures. Often one side of the junction is heavily doped, resulting in a highly asymmetric structure (Fig. 2.3). Reverse bias results in widening the

*The average ionization coefficient is $\alpha_i = 7.03 \times 10^5 \exp(-1.468 \times 10^6 / \mathscr{E})$ in Reference 3. However the value adopted in Reference 6 is used here because it provides an excellent fit to experimental data on breakdown voltage.

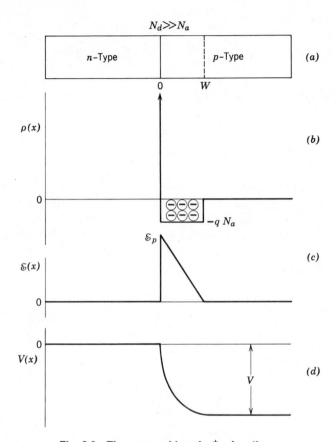

Fig. 2.3 The reverse-biased n^+-p junction.

depletion layer to support this voltage. For an n^+-p junction, depletion layer widening is confined to the p-side. Junction characteristics for the n^+-p diode may be obtained by solving the one-dimensional Poisson equation

$$\frac{d^2V}{dx^2} = -\frac{d\mathcal{E}}{dx} = -\frac{\rho(x)}{\varepsilon\varepsilon_0} = \frac{qN_a}{\varepsilon\varepsilon_0} \qquad (2.24)$$

where $\mathcal{E}$ is the electric field (V/cm), V is the potential (V), $\rho(x)$ is the charge density in the depletion layer (C/cm^3), N_a is the doping level of the p-side (cm^{-3}), ε is the relative permittivity (12 for silicon), and ε_0 is the dielectric constant of free space (8.85×10^{-14} F/cm). Solving, the peak $\mathcal{E}$

field is given by

$$\mathscr{E}_p = \frac{qN_a W}{\varepsilon\varepsilon_0} = \frac{2V}{W} \tag{2.25}$$

The depletion layer width W is given by

$$W = \left(\frac{2\varepsilon\varepsilon_0 V}{qN_a} \right)^{1/2} \tag{2.26}$$

where the voltage supported by the depletion layer V equals the sum of the applied reverse voltage and the contact potential (approximately 1 V). The breakdown voltage of this junction can now be calculated by solving the ionization integral (2.20). This requires knowledge of the depletion layer width W' and the peak electric field $\mathscr{E}_p'$ at the breakdown voltage BV. From (2.25) and (2.26), and ignoring the contact potential,

$$\mathscr{E}_p' = \frac{2BV}{W'} \tag{2.27}$$

and

$$W' = \left(\frac{2\varepsilon\varepsilon_0 BV}{qN_a} \right)^{1/2} \tag{2.28}$$

Combining the ionization integral with (2.27), (2.28), and (2.24), the breakdown voltage is given approximately by [6]

$$BV \cong \frac{\exp(b/\mathscr{E}_p')}{(2a/b)(1-2\mathscr{E}_p'/b)} \frac{\ln\gamma}{\gamma-1} \tag{2.29}$$

This relation can be used together with (2.27) and (2.28) to obtain the BV as a function of N_a. As mentioned earlier, γ is chosen as unity because it provides the best fit to the experimental data.

Figure 2.4 gives the value of BV computed from (2.29), as well as the computer-generated solution of the ionization integral. Note the excellent fit at high breakdown voltages, deteriorating to a fit within 14% for a breakdown voltage of 50 V. Also shown in this figure is the depletion layer width at breakdown, as a function of N_a.

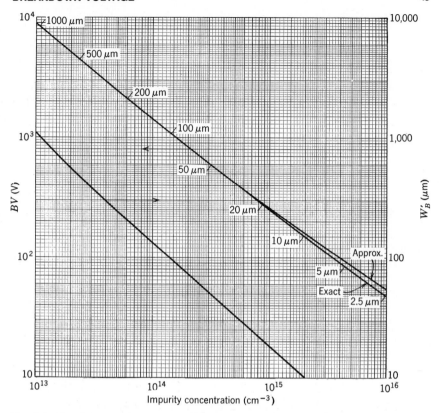

Fig. 2.4 Breakdown voltage for an abrupt junction. Copyright © 1966 by the Institute of Electrical and Electronics Engineers, Inc. Reprinted with permission from Kokosa and Davies [6].

A closed form expression for the depletion layer thickness at breakdown is often useful in the design of rectifier junctions. This can be obtained by using a power series approximation [7] to the ionization coefficient,

$$\alpha_i \cong 1.8 \times 10^{-35}\, \mathscr{E}^7/\text{cm} \qquad (2.30)$$

where $\mathscr{E}$ is the electric field intensity in volts per centimeters. This form allows direct integration of (2.20) to give the breakdown voltage for silicon as

$$BV \cong 5.34 \times 10^{13}\, N_a^{-3/4}\ \text{V} \qquad (2.31)$$

where N_a is in cm^{-3}. The depletion layer thickness at breakdown is given by

$$W' \cong 2.57 \times 10^{-2} BV^{7/6} \mu m \qquad (2.32)$$

where BV is in volts.

The breakdown voltage for abrupt p-n junctions having doping levels of N_a and N_d can also be obtained [8] by using an effective doping concentration N_{eff} such that $1/N_{eff} = 1/N_a + 1/N_d$. All of the equations for the n^+-p junction can be used if this substitution is made for N_a.

2.4.2 The Punched-Through Junction

Figure 2.4 indicates that extremely wide depletion layers are required of diodes with high reverse breakdown voltage. In the forward direction, much of this region is undepleted and can contribute to ohmic drop. To reduce this effect, n^+-p diodes are often designed to ensure that the p-region is fully depleted *before* breakdown occurs. The breakdown voltage of such a *punched-through* (PT) diode is now compared to that of the normal structure, in which punchthrough does not occur.

Consider [9] two abrupt junction n^+-p diodes (Fig. 2.5), both with the same doping level on the p-side. Both are terminated on this side by p^+-regions that serve as ohmic contacts. Let the depletion layer of the long diode (PN) extend to x' at breakdown, and let $\mathscr{E}'_p$ be the peak field for this condition. The punched-through diode is shown with a p-region of width W, where $W/x' = \eta \leqslant 1$. Since $d\mathscr{E}/dx = -qN_a/\varepsilon\varepsilon_0$, the slope of the electric field versus distance characteristic is the same for both diodes.

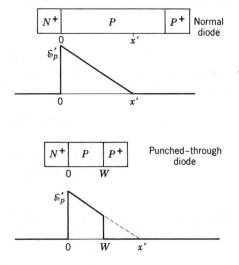

Fig. 2.5   The punched-through diode.

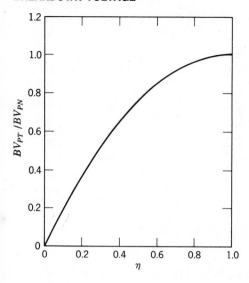

Fig. 2.6 Breakdown voltage ratio for a punched-through diode.

To an approximation, breakdown occurs when each device attains a critical* peak electric field $\mathcal{E}_p'$. In addition, the breakdown voltage is given by the area under the $\mathcal{E}$ versus x curve. Solving,

$$\frac{BV_{PT}}{BV_{PN}} \cong 2\eta - \eta^2 \tag{2.33}$$

Thus the breakdown voltage of the punched-through diode is always less than that of its normal counterpart, as seen in Fig. 2.6. For example, if a half-width depletion layer is used, the BV is reduced by only 25% from that for the normal diode.

An interesting tradeoff can be made along these lines during the design of high voltage junction rectifiers. Consider, for example, a 1700 V device made with a background concentration of 8.0×10^{13} cm^{-3} and having a depletion layer width of 160 μm. This device may be designed to have a lower background concentration, but with a shortened base to form a punched-through structure. A number of alternate designs are given in Table 2.1. Note the shortening of the base width by use of a punched-through structure, until $\eta \cong 0.2$. Beyond this point, there is little gain; in fact, the base width actually increases slightly at very small values of η.

---

*Note that breakdown is not strictly dependent on a critical field. This approximation is reasonable largely because α_n and α_p are strong functions of $\mathcal{E}$, and can be used for comparing closely similar structures.

Table 2.1   Alternate Designs for a 1700 V Diode

N_a (atoms/cm^3)	BV_{PN} (V)	$\dfrac{BV_{PT}}{BV_{PN}}$	η	x' (μm)	W (μm)
8×10^{13}	1700	1	1	160	160
7×10^{13}	1850	.918	.714	180	129
5×10^{13}	2400	.708	.460	240	110
3×10^{13}	3600	.472	.273	380	104
2×10^{13}	5100	.333	.183	560	102
1.5×10^{13}	6500	.262	.140	740	103.6
1×10^{13}	9000	.189	.099	1100	109

2.4.3   The n^+-i-p^+ Junction

The breakdown voltage of the punched-through n^+-p diode can be increased by making the p-region near-intrinsic. This results in the n^+-i-p^+ structure appearing in Fig. 2.7. Let W be the width of the i-region. Since there are no ionized impurities in this region, the electric field is constant, and has a value of $\mathscr{E}'_p$ at breakdown. Then

$$BV = \mathscr{E}'_p W \tag{2.34}$$

Breakdown occurs when

$$\int_0^W \alpha_n \, dx = \frac{\ln \gamma}{\gamma - 1} \tag{2.35}$$

Combining (2.6) and (2.34) with (2.35) gives

$$BV = \frac{Wb}{\ln\{aW(\gamma - 1)/\ln \gamma\}} \tag{2.36}$$

An important advantage of this structure is that BV is now independent of the doping level and only depends on physical dimensions. Equation 2.36 can be further simplified if it is assumed that $\alpha_n = \alpha_p = \alpha_i$. Then

$$BV = \frac{Wb}{\ln aW} \tag{2.37}$$

Once again, the simpler form has been found to provide a better fit to the experimental data than that given by (2.36).

Fig. 2.7 The n^+-i-p^+ diode.

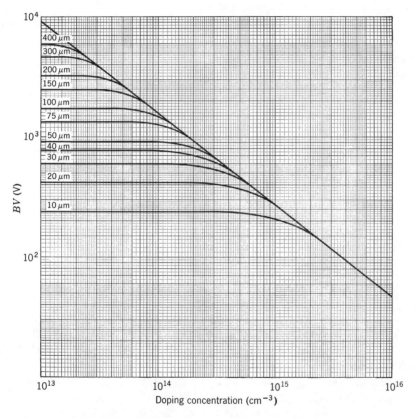

Fig. 2.8 Breakdown voltage for an n^+-p-p^+ diode.

Using this equation and (2.29), we can now synthesize Fig. 2.8 to approximate the breakdown voltage of an n^+-p-p^+ diode as a function of the width of the p-region as well as its doping concentration.

2.4.4   The Linear Junction

The linear junction is formed when the impurity concentration in a semiconductor changes linearly from p-type to n-type over a finite length of material. It can be characterized by a grade constant $\mathcal{C}$, which defines the slope of the impurity concentration and has the dimensions of cm^{-4}. To an approximation, the collector-base junction of a diffused transistor behaves as if linearly graded. Deep diffusions exhibit many of the properties of linearly graded junctions as well.

Figure 2.9 shows a reverse-biased linear junction with a grade constant $\mathcal{C}$. Let the depletion layer extend $W/2$ into each of the p- and n-regions, resulting in the idealized space charge density configuration of Fig. 2.9b. Integration of Poisson's equation results in the electric field $\mathcal{E}$, where

$$\mathcal{E} = \mathcal{E}_p \left[1 - \left(\frac{2x}{W} \right)^2 \right]$$ (2.38)

The peak $\mathcal{E}$ field is given by

$$\mathcal{E}_p = \frac{q\mathcal{C}W^2}{8\varepsilon\varepsilon_0}$$ (2.39)

A second integration gives the reverse voltage across the junction as

$$V = \frac{q\mathcal{C}W^3}{12\varepsilon\varepsilon_0}$$ (2.40)

Again, this voltage is equal to the sum of the applied reverse voltage and the contact potential ($\cong 1$ V). Finally, the peak $\mathcal{E}$ field can be written in terms of the junction voltage by combining (2.39) and (2.40)

$$\mathcal{E}_p = \left(\frac{9q\mathcal{C}V^2}{32\varepsilon\varepsilon_0} \right)^{1/3} = \frac{3V}{2W}$$ (2.41)

The breakdown voltage can be calculated by solving the ionization integral for this junction. Let the width of the depletion layer at breakdown

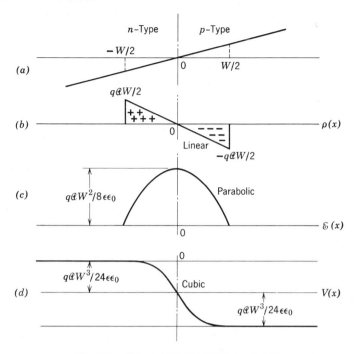

Fig. 2.9 The reverse-biased linear junction.

be $W'/2$ on either side of the junction and the peak field $\mathscr{E}_p'$. Then

$$\mathscr{E} = \mathscr{E}_p'\left[1 - \left(\frac{2x}{W'}\right)^2\right] \tag{2.42}$$

where

$$W' = \left(\frac{8\varepsilon\varepsilon_0\mathscr{E}_p'}{q\,\mathscr{A}}\right)^{1/2} \tag{2.43}$$

Also,

$$\frac{d\mathscr{E}}{dx} = -\mathscr{E}_p'\frac{8x}{W'^2} \tag{2.44}$$

Making these substitutions into (2.20), the ionization integral reduces to

$$\left(\frac{\varepsilon\varepsilon_0}{2q\,\mathscr{A}}\right)^{1/2}\int_0^{\mathscr{E}'}\frac{\alpha_i\,d\mathscr{E}}{\left(\mathscr{E}_p' - \mathscr{E}\right)^{1/2}} = 1 \tag{2.45}$$

This can be solved by setting

$$\alpha_i = A\,\mathscr{E}^7 \qquad \text{and} \qquad A = 1.85 \times 10^{-35} \tag{2.30}$$

Setting $\mathscr{E} = \mathscr{E}'_p \sin^2\theta$ and making this change of variable, (2.45) reduces to

$$\left(\frac{2\varepsilon\varepsilon_0\,\mathscr{E}'_p}{q\mathfrak{a}}\right)^{1/2} A\left(\mathscr{E}'_p\right)^7 \frac{\Gamma(8)\,\Gamma(0.5)}{\Gamma(8.5)} = 1 \tag{2.46}$$

where $\Gamma(y)$ is the gamma function of y. But

$$\mathscr{E}'_p = \left(\frac{9q\,\mathfrak{a}BV^2}{32\,\varepsilon\varepsilon_0}\right)^{1/3} \tag{2.47}$$

Substituting into (2.46) and solving for the various constants, gives (for silicon)

$$BV = 9.17 \times 10^9\,\mathfrak{a}^{-0.4}\ \text{V} \tag{2.48}$$

where $\mathfrak{a}$ is the linear grade constant (cm^{-4}). Figure 2.10 presents this relationship for a linearly graded junction. As expected, the breakdown voltage falls as the junction gradient becomes steeper.

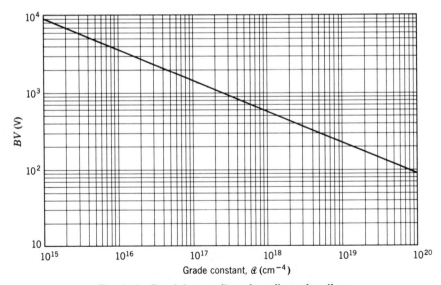

Fig. 2.10 Breakdown voltage for a linear junction.

From (2.40) the breakdown voltage is given by

$$BV = \frac{q\mathbb{Q}\ W'^3}{12\,\varepsilon\varepsilon_0} \qquad (2.49)$$

where W' is the *total* depletion layer width at breakdown. Combining (2.49) with (2.48) gives

$$W' = 2.15 \times 10^{-2}\,BV^{7/6}\,\mu\mathrm{m} \qquad (2.50)$$

It is interesting to note the similarity between (2.32) for the abrupt junction, and (2.50) for the linear junction [10]. In fact, an average expression

$$W' = 2.35 \times 10^{-2}\,BV^{7/6}\,\mu\mathrm{m} \qquad (2.51)$$

holds to $\pm 8\%$ for both abrupt and linear junctions, over the 100–10,000 V range.

2.4.5 The Diffused Junction

Diffusion is the most common technique for fabricating high voltage junctions in power devices. Extremely large junction depths are required to obtain a low impurity gradient concentration, which is necessary for a high breakdown voltage structure. Junction depths typically range from 5 to 25 μm in transistors, and from 25 to 200 μm in high voltage rectifiers and thyristors.

Diffusion is a process [11] by which dopants can be transported into a semiconductor under the influence of a concentration gradient. The resulting impurity profile is of the complementary error function type if the diffusion is made from an *infinite* source, and is given by

$$N(x,t) = N_0\,\mathrm{erfc}\!\left(\frac{x}{2\sqrt{Dt}}\right) - N_B \qquad (2.52)$$

where $N(x,t)$=net impurity concentration at any point in space and time (atoms/cm^3)

N_0=impurity concentration at silicon surface (atoms/cm^3)

D=value of diffusion coefficient for specific diffusion temperature (cm^2/sec)

x=penetration depth (cm)

t=diffusion time (sec)

N_B=background impurity concentration (cm^{-3})

The minus sign is necessary because a junction is obtained by diffusing an impurity into a background concentration of opposite impurity type.

Diffusions made from a *finite* impurity source result in doping profiles of the Gaussian type, given by

$$N(x,t) = N_0 \exp \frac{-x^2}{4Dt} - N_B \tag{2.53}$$

where the terms are defined as previously.

An increasingly common practice for obtaining deep diffusions is to simultaneously diffuse both gallium and aluminum [12] into an n-type background. This results in a junction of the "double complementary error" type, whose surface concentration is equal to that of the more highly soluble dopant (gallium) and whose junction depth is given by the more rapid diffuser (aluminum). In this manner, it is possible to make extremely deep (50–200 μm) junctions in a reasonable diffusion time, and with good electronic transport properties.

For deep diffusions, all these doping profiles are closely approximated by an exponential, which can be written in the compact form

$$N(x) = N_B (e^{-x/\lambda} - 1) \tag{2.54}$$

where N_B is the background concentration (cm^{-3}), x is the distance (μm), and λ is a space constant (μm). Note that (2.54) defines the metallurgical junction at the origin and not at the surface, as in (2.52) and (2.53).

The impurity concentration profile for a diffused junction is essentially exponential on one side of the depletion layer and constant on the other. It follows that outside the depletion layer, there is a constant electric field in the exponentially graded side of the junction and no field in the constant impurity side. The $\mathcal{E}$ field for the entire structure is shown in Fig. 2.11b. The space charge associated with this junction (Fig. 2.11c) is obtained by differentiating the $\mathcal{E}$ field. Note that the regions beyond the depletion layer are quasi-neutral, notwithstanding the presence of this field in the exponentially graded side of the structure.

The properties of the exponentially graded junction may be summarized as follows:

1. The space charge is essentially confined to the depletion layer and to the surface.
2. The regions outside of the depletion layer are quasi-neutral.
3. There is no $\mathcal{E}$ field in the constant concentration side of the junction, outside the depletion layer.
4. There is a constant $\mathcal{E}$ field in the exponential side of the junction. Its direction is such as to oppose the flow of injected minority carriers in

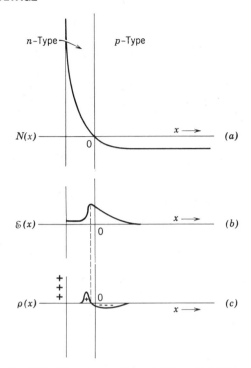

Fig. 2.11 The reverse-biased diffused junction.

this region; that is, it is a retarding field. In addition, there is a slight shift in the effective position of the junction, causing the peak electric field to be located in the heavily doped region, as shown.

The breakdown voltage of a diffused junction may be computed by solving the ionization integral, subject to the appropriate boundary conditions. In general, it is not possible to obtain a closed-form solution for this integral. However numerical solutions of this problem have been obtained [3, 13] with the aid of a digital computer, and they cover the range of breakdown voltages up to 10,000 V (Fig. 2.12). Note that the breakdown voltage of the diffused junction approaches that of a linear junction for small values of grade constant. In addition, it approaches the breakdown voltage of an abrupt junction for large values of a. Finally, the breakdown voltage of a diffused junction is always higher than that for an abrupt junction having the same background concentration.

It has been noted that the breakdown depletion layer width for both abrupt and linear junctions can be approximated by

$$W' = 2.35 \times 10^{-2} BV^{7/6} \, \mu m \qquad (2.51)$$

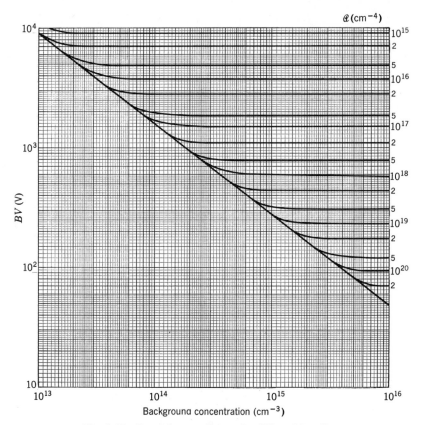

Fig. 2.12   Breakdown voltage of a diffused junction.

where BV is in volts. This equation can be expected to hold for the diffused junction also, since it is bracketed in its properties by the abrupt and linear junction.

The breakdown characteristics of junctions made by the simultaneous diffusion of gallium and aluminum have also been determined [12], and they are very similar to those for diffusion from a single dopant source.

2.5   DEPLETION LAYER CURVATURE

Up to this point, we have only considered the parallel plane structure, obtained by making a junction at a uniform depth, into an infinitely large semiconductor plane. Such a junction is closely approximated over a large percentage of the surface of a typical rectifier diode. In two important situations, however, the approximation of a parallel plane junction breaks down, resulting in curvature of the depletion layer edge. The first occurs at the (finite) boundary of a semiconductor slice, which is always present in a practical situation. The second occurs at the finite boundary of semiconductor diffusions that are made in photomasked structures. Moreover, combinations of these situations are often encountered in practice. In both cases the potential lines become distorted, as do the corresponding electric field lines. Consequently the breakdown voltage in these regions differs from that in the parallel plane (PP) part of the junction.

Consider a reverse-biased n^+-p diode whose depletion layer is curved as in Fig. 2.13a. To a first-order approximation, this diode can be thought of as an infinite number of small devices in parallel, each of which must support the same reverse voltage $V = -\int \mathscr{E}\,dx$. The path of integration associated with the diode at A in Fig. 2.13a is shorter than that for the diode at B, thus higher values of $\mathscr{E}$ will be found in this device. Finally, since calculation of the breakdown voltage involves integration of the ionization coefficient, which is a strong function of $\mathscr{E}$, it is reasonable to expect that the diode at A will have a lower breakdown voltage than the diode at B. By a similar argument, if the depletion layer curvature resembles that in Fig. 2.13b, then $BV_B < BV_A$.

Thus it appears that the breakdown voltage can locally be greater than or less than what is expected in a parallel plane structure, as a result of

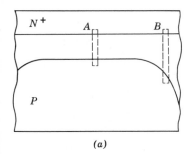

(a)

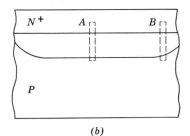

(b)   **Fig. 2.13 Depletion layer curvature.**

depletion layer curvature. Therefore depletion layer curvature can be a significant limitation on reverse diode performance. Alternately, its control can play an important role in the design of high voltage devices.

2.5.1 The Cylindrical Junction

A diffused junction, delineated by means of a photoetched oxide mask, is parallel plane in character except at the edge of the oxide window. Here, owing to lateral diffusion effects [14], the junction takes on a cylindrical shape. As a result, the space charge lines in this region are distorted, resulting in an $\mathcal{E}$ field that differs from that obtained in a parallel plane (PP) structure. To an approximation, the radius of this cylindrical (CY) region is equal to the junction depth; thus distortion effects are more severe in shallow junctions that in deep junctions.

In a cross-sectional view of a cylindrical n^+-p junction of this type (Fig. 2.14), r_j is the radius of the n^+-region, and r_d is the radius of the depletion layer edge at breakdown. Poisson's equation in cylindrical coordinates applies to this situation for $r_j \leqslant r \leqslant r_d$. Thus

$$\frac{1}{r}\frac{d}{dr}\left(r\frac{dV}{dr}\right) = -\frac{1}{r}\frac{d}{dr}(r\mathcal{E}) = \frac{qN_a}{\varepsilon\varepsilon_0} \qquad (2.55)$$

where $\mathcal{E}$ and V are the electric field and potential along a radius vector, and N_a is the doping concentration in the p-region. This equation can be solved, subject to the condition that the electric field be zero at the depletion layer edge. Solving gives

$$\mathcal{E} = \frac{qN_a}{2\varepsilon\varepsilon_0}\left(\frac{r_d^2 - r^2}{r}\right) \qquad (2.56)$$

Further integration of (2.55), including the condition of zero potential at

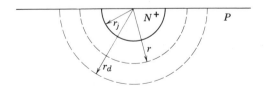

Fig. 2.14   The cylindrical junction.

the metallurgical junction, gives

$$V = \frac{qN_a}{2\varepsilon\varepsilon_0}\left(\frac{r_j^2 - r^2}{2} + r_d^2 \ln\frac{r}{r_j}\right) \tag{2.57}$$

Comparison of these equations with those for the abrupt junction reveals that for the same reverse voltage, the peak electric field is higher for the cylindrical junction. Intuitively, therefore, we can expect the breakdown voltage to be lower in this case. The actual solution for this voltage requires evaluation of the ionization integral for what is essentially a two-dimensional problem, amenable to analysis by computer techniques [15, 16]. The general approach is based on evaluating the ionization integral along paths followed by avalanching hole-electron pairs (Fig. 2.15). Assume that a hole-electron pair is thermally generated at an arbitrary starting point along a field line L_1, so that M_{L1} holes reach point x_1 at the silicon surface. These holes now move along the oxide-silicon interface toward the junction, following the tangential $\mathscr{E}$ field and undergoing ionizing collisions along this path. Consider at x_2 that a hole-electron pair is produced by such a collision. The hole continues to move along the interface, but the electron returns into the depletion layer in the flux tube around the field line L_2. This is turn gives back $(M_{L2} - 1)$ holes to the point x_2 because of avalanching along this field line. The cooperative effect of all current paths of this type must be considered in solving for the multiplication factor. The path of maximum multiplication is *usually* the electric field line that passes through the point of maximum electric field.

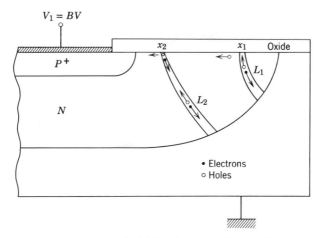

Fig. 2.15 Avalanche multiplication in a cylindrical junction.

Simpler but less accurate techniques for evaluating the ionization integral assume an integration path along the radius vector. Here, too, a closed form solution is not possible. However [17] the seventh power dependence of the ionization coefficient on the electric field results in most of the ionization being confined to the high field regions of the junction depletion layer. Examination of (2.56) shows that for cylindrical junctions, this region is largely confined to small values of r near the boundary of the metallurgical junction. This allows an approximation,

$$\mathscr{E} = \frac{K}{r} \qquad (2.58)$$

for substitution in the ionization integral. This hyperbolic approximation for the electric field results in the extension of the "depletion layer" to infinity. Thus the integration must now be performed from the metallurgical junction boundary r_j to infinity. Carrying out this integration the peak electric field at breakdown is obtained as

$$\mathscr{E}'_{pCY} = \left(\frac{6}{A r_j} \right)^{1/7} \qquad (2.59)$$

where $A = 1.8 \times 10^{-35}$ and r_j is in centimeters. Normalizing to the parallel plane case,

$$\frac{\mathscr{E}'_{pCY}}{\mathscr{E}'_{pPP}} = \left(\frac{3 W'}{4 r_j} \right)^{1/7} \cong \left(\frac{W'}{r_j} \right)^{1/7} \qquad (2.60)$$

Using this relationship between the peak electric field in the cylindrical and parallel plane cases, the breakdown voltage of an abrupt cylindrical junction can be normalized to the corresponding (same background doping) parallel plane junction. This ratio is derived as

$$\frac{BV_{CY}}{BV_{PP}} = \frac{2}{W_c \mathscr{E}'_{pPP}} \left\{ \frac{\mathscr{E}'_{pCY} r_j}{2} + \frac{1}{2} \left(\frac{q N_a}{2 \varepsilon \varepsilon_0} r_j^2 - \mathscr{E}'_{pCY} r_j \right) \left[\ln \left(1 - \frac{2 \varepsilon \varepsilon_0 \mathscr{E}'_{pCY}}{q N_B r_j} \right) \right] \right\} \qquad (2.61)$$

which reduces to

$$\frac{BV_{CY}}{BV_{PP}} = \left\{ \frac{1}{2} \left[\left(\frac{r_j}{W'} \right)^2 + 2 \left(\frac{r_j}{W'} \right)^{6/7} \right] \ln \left[1 + 2 \left(\frac{W'}{r_j} \right)^{8/7} \right] - \left(\frac{r_j}{W'} \right)^{6/7} \right\} \qquad (2.62)$$

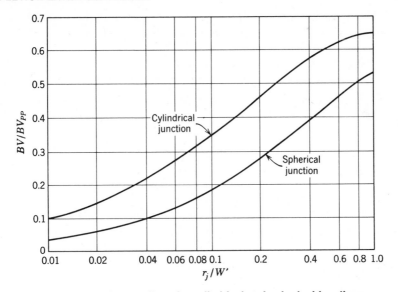

Fig. 2.16 Breakdown voltage for cylindrical and spherical junctions.

Figure 2.16 illustrates this relationship. The agreement between it and computer-generated solutions [18] is extremely good for values of r_j/W' below 0.9, and for background concentrations below $1 \times 10^{16}/\text{cm}^3$. Thus virtually all junctions encountered in semiconductor power devices can be accurately described by this equation.

Computer-aided solutions have also been made [18] for cylindrical junctions of the linear and diffused types. In general, it has been noted that breakdown voltage of a junction is relatively independent of its radius of curvature, provided the electric field is symmetrical with respect to the metallurgical junction. This condition certainly holds for linear junctions, but it is only approximated by diffused junctions.

2.5.2 The Spherical Junction

The curved region where two cylindrical junctions meet can be closely approximated by a spherical junction. The diffusion contour associated with the corner of a rectangular window is of this type. The breakdown voltage of such a junction is calculated by solving Poisson's equation in spherical coordinates,

$$\frac{1}{r^2}\frac{d}{dr}\left(r^2\frac{dV}{dr}\right) = -\frac{1}{r^2}\frac{d}{dr}(r^2 \mathcal{E}) = \frac{qN_a}{\varepsilon\varepsilon_0} \tag{2.63}$$

subject to the boundary conditions of zero electric field at the depletion layer edge (r_d) and zero potential at the metallurgical junction (r_j). Thus

$$\mathcal{E} = \frac{qN_a}{3\varepsilon\varepsilon_0}\left(\frac{r_d^3 - r^3}{r^2}\right) \tag{2.64}$$

and

$$V = \frac{qN_a}{3\varepsilon\varepsilon_0}\left[\frac{r_j^2 - r^2}{2} + r_d^3\left(\frac{1}{r_j} - \frac{1}{r}\right)\right] \tag{2.65}$$

As in the case of cylindrical junctions, the ionization is largely confined to small values of r near the metallurgical junction and the approximation

$$\mathcal{E} = \frac{K}{r^2} \tag{2.66}$$

can be used instead in the ionization integral. Performing the integration along a radius vector from r_j to infinity as before gives the peak electric field at breakdown for spherical junctions:

$$\mathcal{E}'_{pS} = \left(\frac{13}{Ar_j}\right)^{1/7} \tag{2.67}$$

where $A = 1.8 \times 10^{-35}$. Normalizing to the corresponding parallel plane case gives

$$\frac{\mathcal{E}'_{pS}}{\mathcal{E}'_{pPP}} = \left(\frac{13}{8}\frac{W'}{r_j}\right)^{1/7} \tag{2.68}$$

Hence

$$\frac{BV_S}{BV_{PP}} = \frac{2}{W'\mathcal{E}'_{pPP}}\left[\frac{qN_a}{2\varepsilon\varepsilon_0}r_j^2 - \mathcal{E}'_{pS}r_j - \frac{qN_a}{2\varepsilon\varepsilon_0}\left(r_j^3 - \frac{3\varepsilon\varepsilon_0 r_j^2 \mathcal{E}'_{pS}}{qN_a}\right)^{2/3}\right] \tag{2.69}$$

which reduces to

$$\frac{BV_S}{BV_{PP}} = \left(\frac{r_j}{W'}\right)^2 + 2.14\left(\frac{r_j}{W'}\right)^{6/7} - \left[\left(\frac{r_j}{W'}\right)^3 + 3\left(\frac{r_j}{W'}\right)^{13/7}\right]^{2/3} \tag{2.70}$$

A plot of this relationship also appears in Fig. 2.16. Note that for the same radius of curvature, the spherical junction has an even lower breakdown voltage than the cylindrical junction.

Extremely small radii of curvature are encountered in sharp corner regions associated with masked diffusion through rectangular windows. Such corners must always be avoided by rounding them off; often designers of power devices use circular lateral geometry to avoid completely junction breakdown due to spherical curvature.

2.5.3 Diffused Guard Ring Structures

The diffused guard ring can be used to force the curvature of the depletion layer into the shape shown in Fig. 2.13a. The avalanche photodiode provides an excellent example for this scheme, since it allows the use of a shallow diffusion (for efficient photodetection) while avoiding edge breakdown resulting from its sharp curvature.

Figure 2.17 shows an n^+-p structure of this type, which is made by first diffusing a deep n^+-type annular ring, followed by a shallow n^+-circular diffusion. The deep annular diffusion is designed to have a breakdown voltage at its edges that is higher than the BV of the shallow diffusion in its parallel plane region. This can be readily achieved, since the guard ring has a smaller grade constant $\mathcal{C}$ than the shallow n^+-diffusion.

Note that the depletion layer edge of the shallow junction now curves away from the n^+-region. This means that breakdown will occur over the parallel plane section, and uniform avalanching can be achieved.

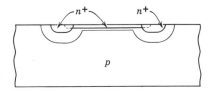

Fig. 2.17 Diffused guard ring structure.

Diffused guard rings can also be incorporated [19] into punched-through structures to limit the field due to depletion layer curvature. To accomplish this, the device is fabricated in the form of a circular p^+-n structure as in Fig. 2.18, with a series of appropriately spaced concentric p^+-ring diffusions. These rings are left floating and are free to adopt any potential during device operation.

The depletion layer is initially associated with the main junction P_0, and extends outward with increasing reverse bias. The spacing between P_0 and

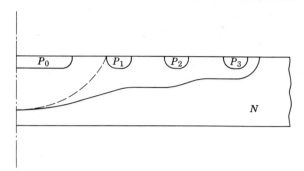

Fig. 2.18   Diffused field-limiting rings.

P_1 is such that punchthrough occurs *before* the avalanche breakdown voltage of the cylindrical junction associated with P_0. Thus the maximum $\mathcal{E}$ field across the main junction P_0 is limited; any further increase in reverse voltage is taken up by P_1 until the depletion layer punches through to P_2, and so on. Ultimately, the device breaks down at the cylindrical junction associated with the last diffused ring.

Consider, in the limit, identically spaced guard rings whose diffusion depth is very large compared to the main diffusion. If BV_{PT} is the punchthrough voltage associated with each spacing, and BV_{CY} is the breakdown voltage of the cylindrical junction associated with the nth guard ring, the breakdown voltage of the device is given by

$$BV \cong nBV_{PT} + BV_{CY} \qquad (2.71)$$

In addition, the voltage adopted by each guard ring varies linearly with the voltage on the main junction, once punchthrough occurs.

At the other extreme, consider extremely shallow guard rings of infinitely small width. These have no effect on the behavior of the main junction and serve only as probes on the depletion layer. For large ring diameters, it can be shown from geometric considerations that the voltage adopted by these rings will vary approximately as the square root of the voltage on the main junction, once punchthrough occurs.

For reasons of economy, it is desirable to have all junctions made in a single diffusion step; this ensures that the practical guard ring structure will fall between the two extremes just described. Furthermore, structures are most commonly limited to a single guard ring to avoid excessively large area. Computer-aided studies of such devices [20] have shown that once punchthrough occurs, the voltage adopted by the guard ring varies as the 0.65th power of the applied reverse voltage.

Figure 2.19 shows the fraction of the parallel plane breakdown voltage (BV_{PP}) that can be achieved by a curved junction having one guard ring,

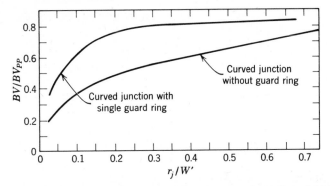

Fig. 2.19   Breakdown voltage of a field-limiting ring structure. Copyright © 1975 by the Institute of Electrical and Electronics Engineers, Inc. Reprinted with permission from Adler et al. [20].

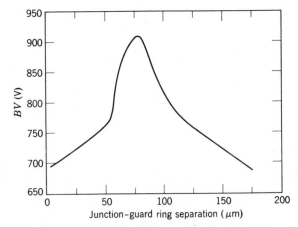

Fig. 2.20   The effect of ring placement. Copyright © 1975 by the Institute of Electrical and Electronics Engineers, Inc. Reprinted with permission from Adler et al. [20].

as a function of r_j/W', where r_j is the radius of curvature of the junction, and W' is the breakdown depletion layer width of a parallel plane device built on the same starting material. For the purpose of comparison, results for a curved junction *without* a guard ring are also shown. Inspection of this figure indicates that the guard ring can improve the breakdown voltage by as much as 80% for small radii of curvature* (i.e., for low

---

*From (2.71) it is seen that even for the ideal case, a single guard ring cannot achieve a 2:1 improvement in breakdown voltage over the junction with no guard ring.

voltage diodes as well as high voltage transistors). However this improvement falls off to only 10% for high voltage diodes, which usually have larger values of r_j / W'.

The breakdown voltage of the guard ring structure has also been shown to be a relatively sensitive function of the separation between the main junction and the guard ring, so that the data of Fig. 2.19 hold only for an optimally spaced structure. The sensitivity to this separation for a typical device is indicated in Fig. 2.20. Thus ring placement must be reasonably precise if full advantage is to be taken of this technique.

2.5.4 Field Plate Structures

The field plate provides an alternate means for control of the depletion layer edge near the surface of a semiconductor, and it is particularly convenient for use with diffused junctions made by photomasking processes. In practice, the field plate is placed on top of the oxide covering the junction and biased with respect to the semiconductor beneath it. As a consequence of charge neutrality, the charge on this plate is balanced by the formation of a space charge layer of opposite polarity type in the semiconductor bulk. Thus positive charges on the plate induce negative charges in the underlying semiconductor, making it more n-type at the surface. In like manner, negative charges on the plate make the underlying semiconductor more p-type. This effective change in conductivity may be used to control the curvature of the depletion layer edge, hence its breakdown voltage, in the region near the surface.

Consider an n^+-p diode like that in Fig. 2.21. Here a field plate is placed on an oxide that is assumed to be ideal and charge-free, and biased with respect to the p-region. The depletion layer edge, for zero voltage (hence zero charge) on the field plate is labeled 0 in this figure. The application to the field plate of negative voltage of successively increasing magnitude, results in the p-region becoming successively more p-type and the depletion layer becoming more tightly curved, as shown by $0, 1, 2$. Thus the BV is reduced from its original value.

In like manner, the application of successively increasing positive voltage on the field plate V_{FP} (hence more positive charge) results in the p-region becoming successively less p-type, thus acquiring increased surface resistivity. The resulting changes in depletion layer edge are labeled $0, 1', 2', 3'$. With each increase in V_{FP}, the curvature is further reduced, resulting in an increased breakdown voltage, approaching that for the parallel plane junction.

If BV volts is applied to the n^+-region (see Fig. 2.21) and V_{FP} to the field plate, the net voltage across the oxide is $BV - V_{FP}$, in the region over

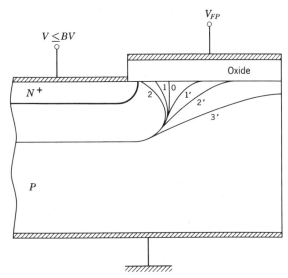

Fig. 2.21   The effect of a field plate on the depletion layer.

the n^+-region. With a thin oxide (relative to the depletion layer width), the maximum $\mathcal{E}$ field concentration occurs in this region. As a result, we can expect, to a crude approximation, that breakdown will occur at a constant value of $BV - V_{FP}$. Thus the breakdown voltage of the junction should be related to the voltage on the field plate by

$$BV \cong V_{FP} + \text{const} \tag{2.72}$$

until the parallel plane breakdown voltage is reached. This has been experimentally observed [21] over wide ranges of values of V_{FP}.

In the foregoing arguments it has been assumed that conductivity changes in the n^+-layer are negligible. In addition, the positive charge applied to the field plate was in all cases insufficient to invert the underlying p-region to n-type. Although it is possible for both these conditions to be violated, values of V_{FP} can be chosen to ensure that this does not happen.

The effect of built-in charges in the oxide is comparable to putting a fixed bias on the field plate. Thermally grown oxides typically have about 10^{11} positive charges per square centimeter. For a $1\,\mu$m thick oxide, this results in an effective bias of about $+5$ V on the field plate, which can be ignored for all practical purposes.

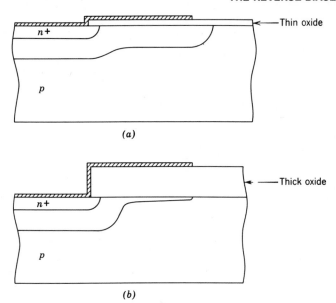

Fig. 2.22   Thin and thick oxides.

In practice, a separate gate voltage supply is avoided by tying the field plate to one region of the junction, (Fig. 2.22*a*). Here, consider an extreme case of a thin oxide on a high voltage junction. By way of example, assume a 1μm oxide and a 100 V junction. The depletion layer width for this junction is about 100 μm, resulting in an average electric field of 10 V/μm in the semiconductor. Since the relative permittivity of silicon dioxide is about 4, whereas that for silicon is 12, the average field in the oxide will be about 30 V/μm. Consequently the depletion layer supports 1000 V in the region below the n^+-diffusion and 970 V in the region covered by the field plate. Thus its width is approximately constant right out to the field plate edge, as shown. All the field concentration region is now out *beyond* the edge of the field plate, and avalanche multiplication preceding breakdown is initiated in this edge region. In addition, since the depletion layer curvature is essentially unchanged, the breakdown voltage is close to that obtained with no field plate and is undesirably sensitive to the effects of extraneous charge migration on the oxide surface.

Figure 2.22*b* shows the other extreme of a field plate on a very thick oxide, say 30 μm.* Again, assume that the voltage supported by the region

*Grown oxides are usually limited to about 1–2 μm thickness before cracking due to differential thermal expansion effects. However pyrolytic oxides and low temperature glasses can be readily deposited in this thickness range.

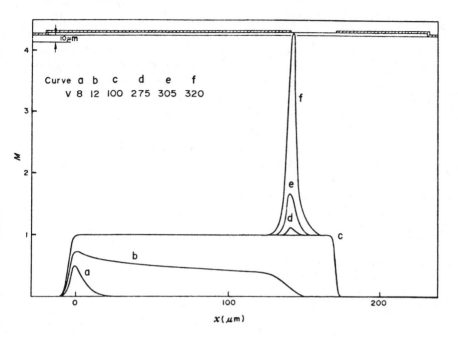

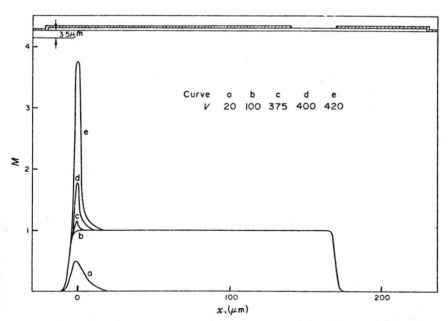

Fig. 2.23 Multiplication in thin and thick oxide field plate structures. Copyright © 1972 by *Solid State Electronics*. Reprinted with permission from Conti and Conti [22].

under the diffusion is 1000 V and that the electric field in the oxide is 30 V/μm. Now, however, 900 V is supported by the depletion layer in the region below the n^+-diffusion, and only 100 V in the region covered by the field plate; thus edge breakdown is eliminated. Yet since the depletion layer curvature near the metallurgical junction is virtually the same as that for no field plate, breakdown occurs in this region at about the same voltage as for a cylindrical junction without a field plate. The field plate is thus ineffective for either extreme situation.

Avalanche multiplication in field plate structures with thin and thick oxides is dramatically illustrated in the experimental data [22] of Fig. 2.23a and 2.23b, respectively. Transparent electrodes were used, and the multiplication factor was determined by measuring the optical emission associated with hole-electron pair recombination in the depletion layer. Figure 2.23a depicts a device with a thin oxide (0.5 μm), that avalanches at the edge of the field plate. The thick oxide (2 μm) structure of Fig. 2.23b is seen to avalanche internally, near the diffused junction boundary. Furthermore, the thin oxide structure has a diffusion depth of 10 μm and a background resistivity of 32 ohm-cm, and it breaks down at 320 V. On the other hand, the thick oxide device has a shallow diffusion (3.5 μm) and a low background resistivity (12 ohm-cm) but breaks down at a somewhat higher voltage (420 V). Thus although neither of these designs is very effective, it is preferable to have avalanching occur near the diffusion edge than at the edge of the field plate.

The ideal field plate structure requires the use of an oxide that is very thin near the diffusion window and thickens outward from the metallurgical junction region. Tapered oxides of this type are not achievable by simple technological means; however stepped oxides have been used to approximate this taper. For example, planar p^+-n diodes with breakdown voltages as high as 900 V have been made [22] with 10 μm diffusion depth, 0.6 μm stepped to 3 μm oxide thickness, and $10^{14}/cm^3$ background concentration. The theoretical value of BV for cylindrical junctions of this type is 500 V without a field plate, whereas the value for the parallel plane junction with the same background concentration is 1400 V. Thus the use of a field plate can result in a significant improvement in breakdown voltage over the circular junction.

2.5.5   Equipotential Rings

Investigations have shown that junctions with field plates often exhibit slowly drifting breakdown characteristics. Much of this problem can be traced to the fact that the oxide has a finite conductivity, so that any charge on the field plate slowly extends out over the surrounding oxide. If

$V(x,t)$ is the voltage at any point on the oxide surface beyond the field plate at any given time, and V_{FP} is the voltage on the field plate, a simple diffusion model gives

$$V(x,t) = V_{FP} \operatorname{erfc}\left(\frac{x}{2\sqrt{t/C_0\rho_\square}} \right) \qquad (2.59)$$

where $\rho_\square$ is the sheet resistance of the oxide surface and C_0 is its capacitance per unit area.

The drift problem is aggravated by ionic contamination of the oxide surface, and by humidity. Phosphorus-doped oxides, often used to control mobile charge caused by sodium contamination, are especially sensitive to moisture, undergoing a change of as much as 3 orders of magnitude in surface conductance from a 40% change in humidity.

In practical structures it is customary to surround the field plate with an equipotential ring, which is tied to the other side of the junction. This may take the form of a metallic plate, tied to a diffused region (Fig. 2.24). Its primary purpose is to ensure that the potential over the insulator surface is established rapidly, thus improving device stability. In addition, it prevents the semiconductor from being inverted in this region. Figure 2.25 shows the potential lines for a typical diode, both with (dotted lines) and without (heavy lines) the equipotential ring. Note that there is little change in the region of high concentration of the electric field near the field plate edge. However the potential lines become more tightly curved near the edge of the equipotential ring, which means that the breakdown voltage can be slightly reduced by its presence. Thus device stability is achieved at the cost of a reduced breakdown voltage. In one experiment [23] the breakdown voltage value was found to be linearly related to the FP–EQR spacing, and it increased at a rate of about 4.3 V/μm.

The FP–EQR structure is ideally suited to monolithic processing techniques. However the technological constraints on this process (oxide masking, open tube diffusions, and junction depths below 20 μm) restrict its use

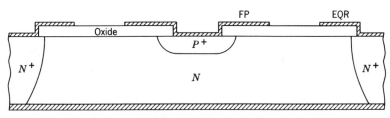

Fig. 2.24 Field plate (FP) with equipotential ring (EQR).

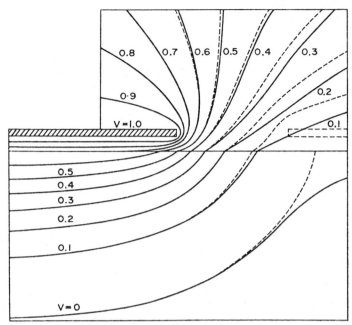

**Fig. 2.25 Map of the equipotential lines with (dashed lines) and without EQR.
Copyright © 1972 by *Solid State Electronics*. Reprinted with permission from Conti
and Conti [22].**

to junctions having a breakdown voltage below 1 kV. For higher voltages,
surface contouring provides an effective alternate technique.

2.6 SURFACE CONTOURING TECHNIQUES

Consider a reverse-biased *p-n* junction, with finite boundaries. On either
side, the depletion layer uncovers bound negative and positive charges,
respectively, resulting in an electric field (Fig. 2.26). Also shown is the
silicon–air boundary, where fringing of the electric field occurs. Since
potential and field lines are orthogonal, the depletion layer in this region is
curved away from the metallurgical junction. The actual extent of curvature depends on the relative doping levels, because charge neutrality
requires that the number of positive charges equal the negative. Thus for a
p^+-n diode, significant curvature of the depletion layer occurs only at the
boundaries of the lightly doped *n*-region.

A second effect to be considered is that of surface charges. For oxidixed
silicon, the surface states typically range in density from $10^{11}/cm^2$ to
$10^{12}/cm^2$, and have a positive charge. With junction coatings such as

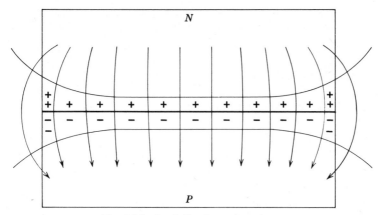

Fig. 2.26 Depletion layer curvature.

silicone resins [24] or low temperature glasses, these charges may be either positive or negative. They further alter the curvature of the depletion layer at the boundaries of a device.

As shown in Section 2.5, depletion layer curvature can result in surface breakdown characteristics that are either better or worse than those of the bulk. Premature surface breakdown is to be avoided in all practical devices, since it causes a concentration of current at the edge and greatly reduces the surge current capability of the junction. Furthermore, the breakdown voltage is now a sensitive function of surface conditions, and not of the bulk resistivity. Finally, even if bulk breakdown occurs before edge breakdown, it is important to reduce the field at the surface, to minimize the effects of ionic contaminant migration.

An important technique [25] for intentionally shaping the depletion layer at the edges consists of contouring the device. By this means, junctions can be designed in which the surface electric field is significantly lower than that in the bulk. An analysis of the effects of surface contouring on the electric field of a reverse-biased *p-n* junction requires the solution of Poisson's equation in two dimensions. The Schottky depletion layer approximation to this equation is

$$\nabla \cdot \mathscr{E} = \frac{q}{\varepsilon \varepsilon_0} \left[N_d\left(x,y\right) - N_a\left(x,y\right) \right] \tag{2.73}$$

where $\mathscr{E}$ is the electric field, q is the electronic charge, N_d, N_a are the donor and acceptor concentrations respectively, ε_0 is the permittivity of free space, and ε is the relative permittivity. Computer-generated solutions of this equation which have been developed by many workers, are now discussed for some simple contours. These contours are not necessarily optimum; however they are relatively easy to implement in practical device manufacture.

2.6.1 Positive-Beveled Junctions

A positive bevel angle is one that results in a junction of decreasing area when going from the heavily doped side to the lightly doped side. Figure 2.27 shows an abrupt p^+-n junction of this type, under a reverse bias of 600 V. Also shown are calculated values of electric field intensity along the surface for different positive bevel angles. For comparison, the calculated value of internal electric field in the bulk is given. The effect of positive bevel angles is noted, as follows:

1. The peak field on the surface is always less than in the bulk, even for a 90° bevel angle (no bevel).
2. The value of the peak electric field falls monotonically as the bevel angle is reduced.
3. The position of the peak electric field shifts away from the metallurgical junction as the bevel angle is reduced, and into the lightly doped side.

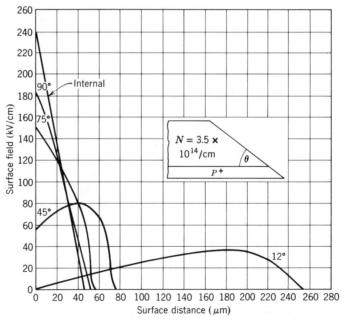

Fig. 2.27 Surface fields in positive beveled structures. Copyright © 1964 by the Institute of Electrical and Electronics Engineers, Inc. Reprinted with permission from Davies and Gentry [25].

Furthermore, calculations for the electric field intensity in a direction from the surface into the bulk reveal that it rises monotonically from its surface value to its bulk value for all positive bevel angles.

The positive bevel angle has no disadvantages, other than the cost of this extra processing step. Consequently, it is the preferred choice for all high voltage junctions. Some situations, however, necessitate the use of a negative bevel angle. The properties of this type of structure are described next.

2.6.2 Negative-Beveled Junctions

It can be shown [25] that a negative bevel angle on a highly asymmetric, uniformly doped, abrupt p^+-n diode causes curvature of the depletion layer edge toward the metallurgical junction at the surface, with a consequent increase of surface electric field over the bulk value. Thus such a structure is not interesting for practical devices.

For a p-n diode that is not highly asymmetric, such as a deep-diffused high voltage structure, however, the surface electric field is increased over the bulk value in the lightly doped region, as before, but the converse is true for the heavily doped side. It is in the heavily doped region, moreover, that the bulk peak electric field is located in diffused junctions (see Section 2.4.5). Thus under certain conditions it is possible to achieve a situation in which junction breakdown does not occur at the surface. This is found to be the case for *extremely small negative bevel angles*.

Figure 2.28 shows the depletion layer for a reverse-biased diffused p-n diode with an exponential doping profile of the type

$$N(x) = N_d(e^{-x/\lambda} - 1) \tag{2.74}$$

where N_d is the background concentration of the starting n-type material,

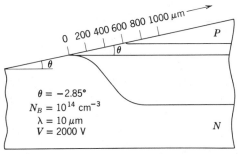

Fig. 2.28 Depletion layer in a diffused junction with a negative bevel. Copyright © 1973 by the Institute of Electrical and Electronics Engineers, Inc. Reprinted with permission from Cornu [26].

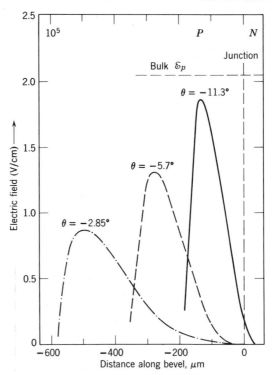

Fig. 2.29 Surface fields in a negative-beveled junction. Copyright © 1973 by the Institute of Electrical and Electronics Engineers, Inc. Reprinted with permission from Cornu [26].

x is the distance from the junction, and λ is a space constant. Figure 2.29 represents the electric field along the surface of this junction for different values of negative bevel angle. The value of the peak electric field, in the bulk, is also indicated in this figure for a reverse bias of 2000 V. Note that for small magnitude bevel angles, the peak electric field at the surface is less than in the bulk and is located in the p-region. In addition, its magnitude at the surface falls monotonically as the magnitude of the negative bevel angle is reduced. Although not shown in this figure, the peak surface electric field exceeds the bulk value, and shifts toward the junction, for larger magnitudes of negative bevel angle [26].

Figure 2.30 gives the magnitude of the peak electric field for these devices, as a function of distance *from the surface into the bulk*. As already noted, $\mathcal{E}_p$ at the surface is less than that in the bulk. However a striking feature here is that this $\mathcal{E}_p$ is not a monotonic function of penetration, but reaches a maximum approximately 25 μm under the surface. Thus these devices exhibit neither surface breakdown nor ideal bulk breakdown! Instead, breakdown occurs at some value of voltage that is below that given by the bulk doping profile, but *internally*. Finally, we note that the magnitude of this internal peak electric field falls as the magnitude of the bevel angle is reduced.

Computer calculations [27] have been made for a wide range of negative-beveled diffused junctions, with surface concentrations from 3×10^{16} to $1 \times 10^{19}/\text{cm}^3$, junction depths from 50 to 200 μm, substrate concentrations from 3×10^{13} to $2 \times 10^{14}/\text{cm}^3$, and ideal values of BV from 962 to 4246 V. The surface of the bevel was assumed to be coated with a material having a dielectric constant of 4 and having no trapped surface charge. It has been shown that all the data can be fitted to within 1% by the single curve (Fig. 2.31). Here, BV_{BVL}/BV_{PP} is the ordinate. The abscissa is a

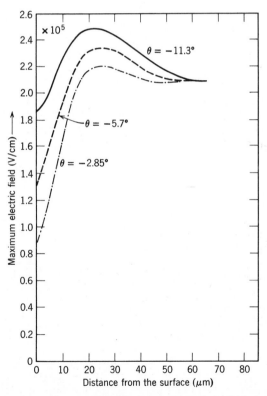

Fig. 2.30   Subsurface fields in a negative-beveled junction. Copyright © 1973 by the Institute of Electrical and Electronics Engineers, Inc. Reprinted with permission from Cornu [26].

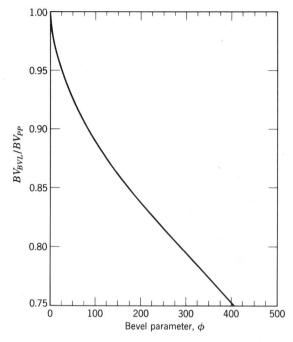

Fig. 2.31 Breakdown voltage of negative-beveled structures. Copyright © 1976 by the Institute of Electrical and Electronics Engineers, Inc. Reprinted with permission from Adler and Temple [27].

normalized bevel parameter ϕ, where

$$\phi = \theta \left(\frac{W_L}{W_H} \right)^2 \tag{2.75}$$

and W_L and W_H are the widths of the depletion layer on the low and high doped sides of the parallel plane part of the junction, respectively.* This figure reveals that the breakdown voltage of the negatively beveled junction approaches its parallel plane value for low values of the bevel parameter. This dictates the use of a deep junction, where (W_L/W_H) approaches unity. In addition, an extremely small bevel angle must be used. The resulting loss of useful area places a severe restriction on the effectiveness of this technique for devices with breakdown voltage above 5 kV.

*Data for parallel plane junctions with these device parameters appear in Table 2.2.

Table 2.2   Properties of Diffused, Parallel Plane *p-n* Junctions

Surface Concentration (cm^{-3})	Substrate Doping, N_d (cm^{-3})	Junction Depth (μm)	W_L (μm)	W_H (μm)	BV_{PP} (V)
3×10^{16}	3×10^{13}	203.2	370.33	60.99	4246
3×10^{16}	5×10^{13}	203.2	228.35	54.76	3092
1×10^{17}	2×10^{14}	88.9	65.28	21.31	1150
1×10^{18}	2×10^{14}	88.9	66.29	16.97	1072
1×10^{19}	2×10^{14}	88.9	67.06	13.84	1024
1×10^{17}	2×10^{14}	50.8	67.06	14.76	1042
1×10^{19}	2×10^{14}	50.8	68.83	9.42	962

Early work with surface contouring, restricted to devices with breakdown voltages below 1000 V, demonstrated that the presence of surface charge was relatively unimportant in determining the shape of the depletion layer boundary. For higher breakdown voltage devices, in the 1–5 kV range, the resistivity of the starting material is significantly higher. As can be expected, the effects of surface charge are important for these devices. Figure 2.32 indicates that a positive surface charge density of $10^{12}/cm^2$ can

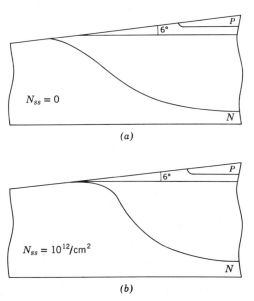

Fig. 2.32   Effect of surface states on depletion layer curvature. Copyright © 1973 by the Institute of Electrical and Electronics Engineers, Inc. Reprinted with permission from Bakowski and Lundstrom [28].

appreciably alter the depletion layer shape for a diode fabricated on a background concentration of $6 \times 10^{13}/\text{cm}^3$. Other parameters of this device are a surface concentration of 3×10^{19} acceptors per cubic centimeter, a junction depth of 100 μm, and a negative bevel angle of 6° [28].

Figure 2.33 presents the surface electric field for this device at a reverse bias of 1760 V, for different values of positive surface state density (N_{ss}). Note that the peak electric field at the surface increases as this density increases. The converse is true also, in that a negative surface charge reduces this electric field. However an excessively high negative surface state density can result in inversion of the underlying n-type semiconductor, with poor breakdown characteristics. Thus the control of this surface charge must be relatively tight, if it is to be beneficial.

2.6.3 Beveled Thyristor Structures

Thyristors are commonly fabricated by making deep, simultaneous, p-type diffusions into both sides of a slice of n-type starting material. Subsequently, an n^+-cathode region is placed on one of these p-regions, usually by alloying. In operation, both these deep junctions must be capable of sustaining a high reverse voltage, necessitating the use of beveling. Thus one junction ordinarily has a positive bevel angle, and the other has a negative angle. Since negative angles are effective only if extremely small, a single bevel for the entire slice results in considerable loss of active device area. As a consequence, it is customary to use double bevel angles; that is, one junction has a large positive bevel angle (30–60°), and the other has a small negative one. Such a structure (Fig. 2.24a) results in area utilization considerably superior to what would be obtained if the device had a single bevel with angles ±6°, for example.

Considerably better area utilization can be achieved if the device is made with positive bevel angles for both high voltage junctions [29], as in Fig. 2.34b, since relatively large bevel angles can now be used for both junctions. Figure 2.35 shows a slice of this type, with two p-type diffusions and 30° positive bevels.* In a practical device design, the depletion layer edge penetrates well over half the width of the n-region when either junction is reverse biased, as shown here for a voltage of 4 kV: in Fig. 2.35, the depletion layer extends beyond the first bevel region and into the second. Thus beyond the point K, the applied voltage must be distributed over a greatly shortened distance along the bevel. The effect of this on the surface electric field is illustrated in Fig. 2.36. Included for comparison is the situation for a single positive bevel angle of 30°. Although the double

---

*The intersection of these two bevels is not sharp, for practical reasons.

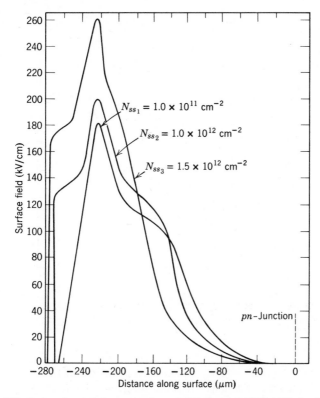

Fig. 2.33 Effect of surface states on surface field. Copyright © 1973 by the Institute of Electrical and Electronics Engineers, Inc. Reprinted with permission from Bakowski and Lundstrom [28].

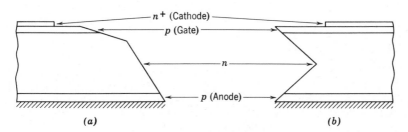

Fig. 2.34 Thyristor structures.

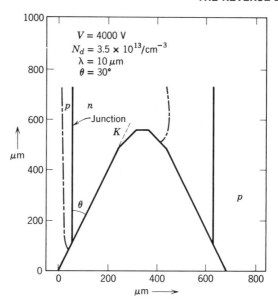

Fig. 2.35 Double positive-beveled thyristor. Copyright © 1974 by the Institute of Electrical and Electronics Engineers, Inc. Reprinted with permission from Cornu et al. [29].

positive bevel is thus somewhat poorer in its ability to lower the electric field at the surface, it has many advantages over the thyristor structure with one positive and one negative bevel angle. Thus it results in excellent area utilization and does not have a subsurface peak electric field that exceeds the bulk value. As a technique, double beveling becomes more attractive as the operating voltage of thyristors is extended into the 4–10 kV range, since useful negative bevel angles for these devices are less than 1°.

2.6.4 Beveled p^+-n-n^+ Structures

The most effective technique for control of the electric field in p^+-n-n^+ structures is to positively bevel the p^+-n junction. This causes the potential lines to spread away from this junction and reduces the surface electric field to values below the bulk. Once the depletion layer reaches through to the n^+-side, however, these potential lines crowd toward the n-n^+ interface, and the peak field moves across to this side. Figure 2.37 shows a structure of this type, with a 50 μm thick n-region. The surface field is also shown for a reverse bias of 500 V. Note that the location of its peak value shifts from the p^+-n to the n-n^+ side as the bevel angle is lowered from

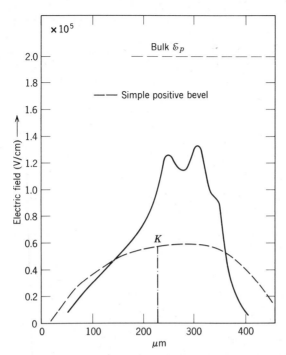

Fig. 2.36 Surface field in a double positive-beveled structure. Copyright © 1974 by the Institute of Electrical and Electronics Engineers, Inc. Reprinted with permission from Cornu et al. [29].

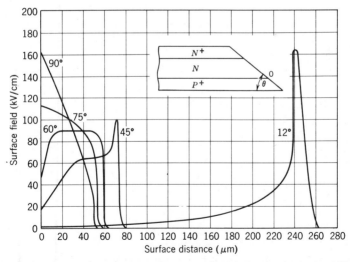

Fig. 2.37 The beveled p^+-n-n^+ diode. Copyright © 1964 by the Institute of Electrical and Electronics Engineers, Inc. Reprinted with permission from Davies and Gentry [25].

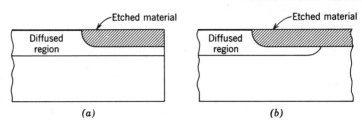

Fig. 2.38   Simple etch contours.

90° to 12°. In addition, the magnitude of the peak electric field first falls, then rises, as this angle is reduced.

2.6.5   Simple Etch Contours

Simple etch contours have been used primarily to remove edge damage incurred during separation of devices on a silicon slice. Highly oxidizing etches ($HNO_3 : HF : CH_3COOH = 8 : 1 : 1$ is a typical formulation) are preferred, since they leave a relatively thick chemical oxide on the etched surface. Such etches, in the form of a deep moat, can be used to form a positive bevel angle at the collector-base junction of high voltage transistors. By this technique, and precise control of the moat depth, breakdown voltages of up to 80% of the parallel plane value have been achieved.

Recently etch contours of the type appearing in Fig. 2.38 have been proposed [30]. By etching into the depleted region, these contours force the potential lines on the heavily doped side to spread out, thus reducing the electric field intensity along the surface of the junction. Structures with both plane and cylindrical junctions have been etched in this manner and have achieved breakdown voltages close to the ideal parallel plane value.

Etch contouring is potentially more economical than beveling techniques and therefore is very attractive. The technique is extremely sensitive to etch depth, however, and precise control of this operation is necessary for its successful implementation.

2.7   INSTABILITIES

A number of instabilities can arise in a reverse-biased junction diode that is operated in the vicinity of its avalanche breakdown voltage. Some of these result in a time-dependent drift in the value of BV; others cause negative resistance effects and may lead to destruction of the device. One form of instability, which results in fluctuation phenomena during the initiation of one or more microplasmas, can be explained [31] by the alternate capture and emission of charge at deep levels in the depletion layer. At the onset of breakdown, the mobile hole-electron concentrations increase rapidly and may become as high as 10^{14}–$10^{16}/cm^3$. As a result, these centers can now have significant recombination properties.

In the n^+-p diode in Fig. 2.39, reverse current flow is largely due to hot holes in the depletion layer on the p-side. At the onset of avalanche, these holes are captured at recombination sites and cause widening of the space charge layer. Thus the breakdown voltage rises, the avalanche is quenched, and the centers discharge until the process repeats. The period of this fluctuation is τ, the average time for a recombination center to capture a hole. If v_{th} is the thermal velocity, σ_p is the capture cross section, and p is the hole concentration, then $\tau \cong 1/v_{th}\sigma_p p$.

A recombination level of about 0.3 eV below the conduction band has been identified [32] as the primary center involved in this process. Its small capture cross section ($\cong 10^{-20}$ cm^2) suggests a doubly charged state, possibly due to divacancies, which are commonly present in the vicinity of crystal damage. A analogous process exists for p^+-n diodes, where the reverse current is predominantly due to the injection of hot electrons into the depletion layer on the n-side.

If the junction is oxide coated, the injection of hot carriers into the depletion layer can be accompanied by trapping effects in this oxide [33].

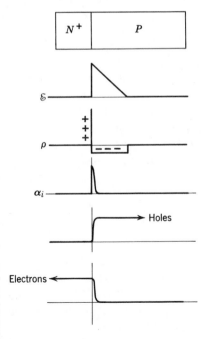

Fig. 2.39 The n^+-p junction at breakdown.

This is because the energy associated with hot holes and electrons during avalanche breakdown is approximately equal to the barrier height between silicon and silicon dioxide at both the valence and conduction band edges ($\cong 3.2$ eV). As a consequence, some of the hot carriers moving in the high field region near the interface have enough energy to surmount this barrier and be injected into the oxide, where they may become trapped. In p^+-n diodes, this trapping (of hot electrons) is assisted by the presence of an electric field created by the built-in positive charge in the oxide layer.

For the n^+-p diode of Fig. 2.39, hole trapping occurs in the oxide over the depletion layer on the p-side. This induces an equal and opposite negative charge in the underlying p-type material, raising its effective resistivity. The breakdown voltage of the junction correspondingly increases until charge relaxation occurs. This may take many days at room temperature, and the time required depends very critically on the quality of the surface oxide, especially on its water content. As a consequence, the breakdown voltage of this n^+-p diode increases with each successive avalanching. This phenomenon, known as *junction walk-out* [34], is also present in p^+-n diodes, where the increase in breakdown voltage is caused by hot electron injection in the oxide over the n-region.

The problem of junction walk-out cannot be eliminated, but it can be reduced by careful processing. In particular, long, low temperature bakeout (24 hr at 250°C), which reduces the water content in the oxide, has been found to be beneficial. Stress relief of the overlying metal film is also effective in some situations.

A negative resistance effect, present in avalanching n^+-ν-p^+ diodes, is also caused by the transport of mobile carriers through a depletion layer [35,36]. Here, however, the depletion layer width is fixed by that of the high resistivity region, and the net result is quite different.

Consider the n^+-ν-p^+ diode in Fig. 2.40, together with its ρ and $\mathscr{E}$ field configurations. Prior to avalanche, the peak electric field is at the ν-p^+ junction; ionization occurs here, and reverse current flow is primarily due to electron transport in the ν-region (i.e., single injection). At the onset of avalanche the electron concentration rapidly increases, distorting the space charge density and causing the electric field to increase at the n^+-ν interface (Fig. 2.37d, 2.37e). Eventually impact ionization occurs at this junction also and results in hole injection into the high resistivity region as well (i.e., double injection). The ν-region is now heavily conductivity modulated, and less voltage is required to support the same current flow through it. This produces a gross negative resistance at breakdown, of the type described in Section 1.4 (Fig. 1.6). This CCNR characteristic results in filamentary current flow, which can lead to mesoplasma formation during avalanche breakdown, seriously impairing device reliability.

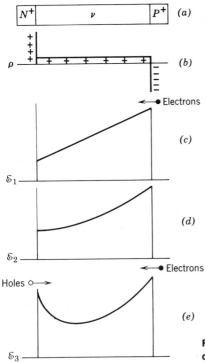

Fig. 2.40 The p^+-ν-n^+ junction at break-down.

The p^+-n-p^+ punchthrough structure has been used [37] to avoid the problems of filamentary conduction during avalanche breakdown. In this structure, current flow is by single carrier injection (see Section 1.3), once the punchthrough voltage is exceeded. Thus negative resistance effects are not present during breakdown, and the whole area of the device is effective in dissipating power during this mode of operation. Note, however, that the power dissipation in the OFF condition is considerably higher than that of a junction diode. As a result, a device of this type is indicated only when high breakdown surge capability is important.

The p^+-n-p^+ punchthrough structure exhibits device characteristics similar to those of two diodes placed back to back. However normal diode-like characteristics can be obtained [38] by building a p^+-ν-n^+ diode in shunt with a p^+-ν-p^+ structure. This is done by placing an n^+-diffusion as well as a p^+-contact at one end (Fig. 2.41). Under reverse bias conditions, indicated in the figure, the device behaves like a p^+-ν-p^+ transistor with a wide base region, in shunt with a reverse-biased p^+-ν-n^+ diode. Once punchthrough occurs, current flows primarily through the p^+-ν-p^+ device and is space charge limited. Experimental data have

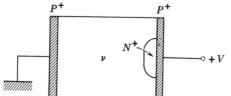

Fig. 2.41 The p^+-n-p^+ punched-through structure.

registered an improvement in peak power dissipation by as much as a factor of 5 at a 200°C ambient temperature, over the conventional n^+-ν-p^+ structure.

2.7.1 Mesoplasmas and Second Breakdown

During avalanche breakdown, localized heating at microplasma sites leads to the thermal generation of mobile carriers. This causes the collapse of the electric field required to sustain the avalanche. Additionally, the mesoplasma begins to form on the lightly doped region of the diode, since the intrinsic temperature is lower on this side. The general characteristics are very similar to those described for bulk material in Chapter 1, provided the appropriate value of intrinsic temperature is used. Experimental verification of this quenching has been established [39] by direct observation of these processes in surface-oriented silicon-on-sapphire diodes.

Mesoplasma formation is aggravated by the presence of deep levels in the depletion layer. As shown in Section 2.1, these act as generation centers with thermal generation current density inversely proportional to the space charge generation lifetime. This is especially significant for centers located near the middle of the energy gap. Here, the space charge lifetime is given by

$$\tau_{sc} = \frac{\tau_{p0} + \tau_{n0}}{2} \tag{2.76}$$

For asymmetric centers, however, the lifetime is considerably larger and is given by [see (1.19)]

$$\tau_{sc} = \frac{\tau_{p0}}{2}\left[\exp\left(\frac{E_r - E_i}{kT}\right)\right] + \frac{\tau_{n0}}{2}\left[\exp\left(\frac{E_i - E_r}{kT}\right)\right] \tag{2.77}$$

where $E_r - E_i$ is the displacement of the center above the middle of the energy gap and may take on positive or negative values. Thus the electronic properties of the contaminant that results in short lifetime are of significance. Alternately, the choice of impurity may be dictated by this consideration in high speed devices where deep levels are intentionally introduced (see Chapter 6). In general, the reduction of lifetime by deep levels that are symmetrically situated in the energy gap $(E_r - E_i \cong 0)$ is accompanied by a comparable increase in the space charge generation current density. On the other hand, this is not true for asymmetric deep levels $(E_r - E_i \neq 0)$. Mesoplasmas thus can be avoided (or delayed) by using device structures that do not lead to microplasma formation, by careful processing so as to maintain long lifetimes, and by using suitable lifetime killers in high speed devices. In addition, contact materials having high eutectic melting temperatures with silicon can be used to improve the second breakdown characteristics of junction diodes.

2.8 REFERENCES

1. C. T. Sah et al., "Carrier Generation and Recombination in *P-N* Junctions and *P-N* Junction Characteristics," *Proc. IRE*, **45**, No. 9, pp. 1228–1243 (1957).

2. J. M. Moll, *Physics of Semiconductors*, McGraw Hill Book Co., New York, 1964.

3. R. Van Overstraeten and H. DeMan, "Measurements of the Ionization Rates in Diffused Silicon *p-n* Junctions," *Solid State Electron.* **13**, No. 5, pp. 583–608 (1970).

4. R. J. McIntyre, "Multiplication Noise in Uniform Avalanche Diodes," *IEEE Trans. Electron Devices*, **ED-13**, No. 1, pp. 164–168 (1966).

5. N. R. Howard, "Avalanche Multiplication in Silicon Junctions," *J. Electron. Control*, **13**, pp. 537–544 (1962).

6. R. A. Kokosa and R. L. Davies, "Avalanche Breakdown of Diffused Silicon *p-n* Junctions," *IEEE Trans. Electron Devices*, **ED-13**, No. 12, pp. 874–881 (1966).

7. W. Fulop, "Calculation of Avalanche Breakdown of Silicon *p-n* Junctions," *Solid State Electron.*, **10**, No. 1, pp. 39–43 (1967).

8. P. Brook, "The Breakdown Voltage of Double-Sided *p-n* Junctions," *IEEE Trans. Electron Devices*, **ED-21**, No. 11, pp. 730–731 (1974).

9. Y. C. Kao, "The Design of High Voltage High-Power Silicon Junction Rectifiers," *IEEE Trans. Electron Devices*, **ED-17**, No. 9, pp. 657–660 (1970).

10. R. W. Warner, Jr., "Avalanche Breakdown in Silicon Diffused Junctions," *Solid State Electron.*, **15**, No. 12, pp. 1303–1318 (1972).

11. S. K. Ghandhi, *The Theory and Practice of Microelectronics*, John Wiley & Sons, New York, 1968.

12. M. Bakowski and I. Lundstrom, "Calculation of Avalanche Breakdown Voltage and Depletion Layer Thickness in a *p-n* Junction with a Double Error Function Doping Profile," *Solid State Electron.*, **16**, pp. 611–616 (1973).

13. D. P. Kennedy and R. R. O'Brien, "Avalanche Breakdown Characteristics for a Diffused *p-n* Junction," *IRE Trans. Electron Devices*, **ED-9**, No. 6, pp. 478–483 (1962).

14. D. P. Kennedy and R. R. O'Brien, "Analysis of the Impurity Atom Distribution Near the Diffusion Mask for a Planar *p-n* Junction," *IBM J. Res. Dev.*, **9**, No. 3, pp. 179–186 (1965).

15. C. Bulucea et al., "Surface Breakdown in Silicon Planar Junctions—A Computer-Aided Experimental Determination of the Critical Field," *Solid State Electron.*, **17**, No. 9, pp. 881–888 (1974).

16. V. A. K. Temple et al., "Calculation of the Diffusion Curvature Related Avalanche Breakdown in High Voltage Planar *p-n* Junctions," *IEEE Trans. Electron Devices*, **ED-22**, No. 10, pp. 910–915 (1975).

17. B. J. Baliga and S. K. Ghandhi, "Analytical Solutions for the Breakdown Voltage of Abrupt Cylindrical and Spherical Junctions," *Solid State Electron.*, **19**, No. 9, pp. 739–744 (1976).

18. S. M. Sze and G. Gibbons, "Effect of Junction Curvature on Breakdown Voltage in Semiconductors," *Solid State Electron.*, **9**, pp. 831–845 (1966).

19. Y. C. Kao and E. D. Wolley, "High-Voltage Planar *p-n* Junctions," *Proc. IEEE*, **55**, No. 8, pp. 1409–1414 (1967).

20. M. S. Adler et al., "Theory and Breakdown Voltage for Planar Devices with a Single Field Limiting Ring," *IEEE Trans. Electron Dev.*, **ED-24**, No. 2, pp. 107–113 (1977).

21. A. S. Grove et al., "Effect of Surface Fields on the Breakdown Voltage of Planar Silicon *p-n* Junctions," *IEEE Trans. Electron Devices*, **ED-14**, No. 3, pp. 157–162 (1967).

22. F. Conti and M. Conti, "Surface Breakdown in Silicon Planar Diodes Equipped with Field Plate," *Solid State Electron.*, **15**, pp. 95–105 (1972).

23. D. S. Zoroglu and L. E. Clark, "Design Considerations for High Voltage Overlap Annular Diodes," *IEEE Trans. Electron Devices*, **ED-19**, No. 1, pp. 4–8 (1972).

24. M. Conti and F. Tegnani, "Electrical Properties of Silicone Films on Silicon," *J. Electrochem. Soc.*, **116**, No. 3, pp. 377–380 (1969).

25. R. L. Davies and F. E. Gentry, "Control of Electric Field at the Surface of *p-n* Junctions," *IEEE Trans. Electron Devices*, **ED-11**, No. 7, pp. 313–323 (1964).

26. J. Cornu, "Field Distribution Near the Surface of Beveled *p-n* Junctions in High Voltage Devices," *IEEE Trans. Electron Devices*, **ED-20**, No. 4, pp. 347–352 (1973).

27. M. S. Adler and V. A. K. Temple, "A General Method for Predicting the Avalanche Breakdown Voltage of Negative Bevelled Devices," *IEEE Trans. Electron Devices*, **ED-23**, No. 8, pp. 956–960 (1976).

28. M. Bakowski and K. I. Lundstrom, "Depletion Layer Characteristics at the Surface of Beveled High Voltage *p-n* Junctions," *IEEE Trans. Electron Devices*, **ED-20**, No. 6, pp. 550–563 (1973).

29. J. Cornu et al., "Double Positive Beveling: A Better Edge Contour for High Voltage Devices," *IEEE Trans. Electron Devices*, **ED-21**, No. 3, pp. 181–183 (1974).

30. V. A. K. Temple and M. S. Adler, "The Theory and Application of a Simple Etch Contour for Near-Ideal Breakdown Voltage in Plane and Planar *P-N* Junctions," *IEEE Trans. Electron Devices*, **ED-23**, No. 8, pp. 950–955 (1976).

31. R. H. Haitz, "Variation of Junction Breakdown Voltage by Charge Trapping," *Phys. Rev.*, **138**, No. 1A, pp. A260–A267 (5 April 1965).

32. K. I. Nuttall and M. W. Nield, "An Investigation into the Behavior of Trapping Centers in Microplasmas," *Solid State Electron.*, **18**, No. 1, pp. 13–23 (1975).

33. E. H. Nicollian et al., "Avalanche Injection Currents and Charging Phenomena in Thermal SiO$_2$," *Appl. Phys. Lett.*, **15**, No. 6, pp. 174–177 (1969).

34. J. F. Schenck, "Burst Noise and Walkout in Degraded Silicon Devices," *Proc. 6th Annual Reliability Physics Symposium*, IEEE Electron Devices Group, and IEEE Reliability Group, Las Vegas, Nevada, pp. 31–39 (1967).

35. H. Egawa, "Avalanche Characteristics and Failure Mechanism of High Voltage Diodes," *IEEE Trans. Electron Devices*, **ED-13**, No. 11, pp. 754–758 (1966).

36. M. W. Muller and H. Guckel, "Negative Resistance and Filamentary Currents in Avalanching Silicon p^+-i-n^+ Junctions," *IEEE Trans. Electron Devices*, **ED-15**, No. 8, pp. 560–568 (1968).

37. P. J. Kannam, "Design Concepts of High Energy Punchthrough Structures," *IEEE Trans. Electron Devices*, **ED-23**, No. 8, pp. 879–882 (1976).

38. J. H. King and J. Philips, "Power Absorption Capability of Punch Through Devices," *Proc. IEEE*, **55**, No. 8, pp. 1361–1365 (1967).

39. R. A Sunshine and M. A. Lampert, "Second Breakdown Phenomema in Avalanching Silicon-on-Sapphire Diodes," *IEEE Trans. Electron Devices*, **ED-19**, No. 7, pp. 873–885 (1972).

2.9 PROBLEMS

1. Compare the relative space charge generation leakage currents of two diodes, one doped with gold and the other with a fictitious deep level with similar electronic properties, except that it is located at 0.22 eV above the valence band. Use the parameters provided in Problem 2, Chapter 1, and assume room temperature.

2. Repeat Problem 1 at 125°C.

3. Develop a graph similar to that of Fig. 2.2 for the variation of the diffusion-limited current density of a p^+-n diode. Ignore the temperature dependence of mobility, and assume that the n-region is fully ionized.

4. A p^+-diffusion is made into a background concentration of 10^{14}/cm^3, resulting in a space constant of 10 μm. What is the junction depth, assuming a surface concentration of 10^{18}/cm^3?

5. A p^+-n diode is made by diffusing into a background concentration of 10^{14}/cm^3. Sketch BV as a function of λ and $\mathcal{C}$ (λ values from 2 to 20 μm), assuming that the junction is (a) abrupt, (b) linear, and (c) diffused. Hence determine the nature of the error involved in using abrupt and linear junction theories for BV.

6. Repeat Problem 5 for background concentrations of 10^{13}/cm^3 and 10^{15}/cm^3.

7. Compute the peak electric field at breakdown for junctions described by Fig. 2.4. Hence verify the footnote to Section 2.4.2.

3

The Forward-Biased Diode

CONTENTS

A POWER SEMICONDUCTOR device must be capable of carrying a high current when operated in its conducting or ON state. To keep the amount of power dissipated to a minimum, this current-carrying capacity should be obtained in conjunction with a low forward voltage drop. Wherever possible, unnecessary ohmic drops must be minimized by fabricating devices on heavily doped epitaxial substrates. Often these substrates are thinned by lapping before mounting on the header. Ohmic drop is neglected from consideration in the following sections, since it is not involved in establishing junction characteristics. However it sets an ultimate limit on the current that can be carried by the device.

This chapter discusses the forward conduction characteristics of junctions used in power rectifiers, transistors, and thyristors. All these junctions incorporate a built-in potential barrier that inhibits the flow of carriers between the p- and n-regions. With forward-applied voltage, the height of this potential barrier is reduced, resulting in a more ready flow of carriers across the junction. It must be emphasized, however, that a finite barrier height exists at all times; in its absence, carriers would be transported across the junction without any externally applied force, resulting in infinite current flow.*

Carrier flow across a p-n junction is by both diffusion and drift. In equilibrium, there is an exact balance between these components of current. Under forward bias conditions, the drift component is reduced, resulting in a net current flow in the forward direction, and against the direction of the electric field.

3.1 THE ABRUPT JUNCTION

Both alloyed junctions and shallow diffused junctions can be considered to have an abrupt doping profile. Thus the emitter-base junction of a transistor as well as the cathode-gate junction of a thyristor can be treated as abrupt. In general, these junctions have highly asymmetrical doping characteristics and are often considered to be one-sided.

3.1.1 Forward Injection

Injection in a forward-biased diode can be classified as ultra-low level, low level, or high level, depending on the assumptions appropriate to the device analysis. The central assumption of *ultra-low level injection* is that recombination occurs primarily within the space charge region, because of the presence of deep impurity levels [1].

*Provided ohmic drops in the p- and n-regions are neglected.

Assume that the recombination centers are located at the middle of the energy gap, so that $E_r = E_i$. Assume further that $\tau_{p0} = \tau_{n0} = \tau$. Then from the Shockley-Read-Hall theory, the recombination rate U is given by

$$U = \frac{np - n_i^2}{\tau(n + p + 2n_i)} \tag{3.1}$$

Because of the assumption of quasi-equilibrium, the minority carrier concentration at the edge of the depletion layer on the p-side is given by

$$n_P(0) = \bar{n}_P e^{qV_A/kT} \tag{3.2}$$

where $\bar{n}_P$ is the equilibrium value and V_A is the applied voltage. Writing the majority carrier concentration at this edge as $p_P(0) \cong \bar{p}_P$, the pn product at the depletion layer edge on the p-side is given by

$$pn = n_i^2 e^{qV_A/kT} \tag{3.3}$$

In like manner, the pn product at the depletion layer edge on the n-side is also given by (3.3). Assume that this product is approximately constant throughout the depletion layer.

From (3.1) it is seen that for any given forward bias, the recombination rate is a maximum when $(p + n)$ is a minimum, subject to the condition of a constant pn product given by (3.3). Thus U is maximized when $p = n$. Substituting in (3.1), we have

$$U_{\max} = \frac{n_i}{2\tau} \frac{e^{qV_A/kT} - 1}{e^{qV_A/2kT} + 1} \tag{3.4}$$

$$= \frac{n_i}{2\tau}(e^{qV_A/2kT} - 1) \tag{3.5}$$

The forward current flow is against the direction of the electric field, as mentioned earlier. Thus for an n-p diode in which the electric field is in the positive direction $(n \rightarrow p)$, the forward current density due to recombination in the depletion layer of width w is given by

$$J_{F,\text{rec}} = -\frac{qn_i w}{2\tau}(e^{qV_A/2kT} - 1) \tag{3.6}$$

We now consider forward bias conditions under which recombination occurs predominantly in the neutral regions beyond the depletion layer. For *low level injection* the change in majority carrier concentration due to the injected minority carriers is so small in this region that its effect can be neglected. Thus for a uniformly doped n^+-p diode there is no electric field in the p-region under conditions of current flow, and injected carriers move only by diffusion. For this situation, assuming quasi-equilibrium and an infinitely long p-region, the minority carrier concentration is given by

$$n_P = n_P(0) e^{-x/L_n}$$ (3.7)

where L_n is the electron diffusion length and $n_P(0) = \bar{n}_P e^{qV_A/kT}$. Since carrier flow is entirely by diffusion, the electron current density is

$$J_n = -qD_n \left(\frac{dn_P}{dx} \right)_{x=0}$$ (3.8)

where D_n is the diffusion constant for electrons. In an n^+-p diode, the entire forward current is carried by this term. Solving, we write

$$J_{F,\text{diff}} = J_n = -\frac{qD_n \bar{n}_P}{L_n} (e^{qV_A/kT} - 1)$$ (3.9)

where V_A is the applied voltage across the diode (positive for forward bias), and ohmic drops due to body resistance are ignored. If the p-region is of finite length W, it can be shown [2] that (3.9) must be modified to

$$J_{F,\text{diff}} = -\frac{qD_n \bar{n}_P}{L_n \tanh(W/L_n)} (e^{qV_A/kT} - 1)$$ (3.10)

In (3.9) and (3.10), the minus sign indicates that current flow is against the direction of the electric field.

At *high injection levels* there is a significant alteration of the majority carrier concentration outside the depletion layer, which gives rise to an electric field. Thus carrier motion is influenced by both diffusion and drift in this region. The presence of the electric field results in a voltage drop across this region. Hence only a fraction of the applied voltage appears across the junction, even if ohmic drops due to body resistance are ignored.

Consider, as before, an abrupt n^+-p junction with a wide, uniformly doped p region. The minority and majority carrier concentrations in this region are given in Fig. 3.1 for high level injection conditions. For approximate charge neutrality, it is assumed that $dp_P/dx \cong dn_P/dx$. Furthermore, assuming that quasi-equilibrium conditions still apply, $D_p = (kT/q)\mu_p$.

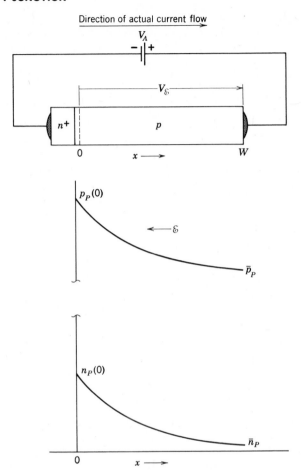

Fig. 3.1 High level injection.

For the p-region

$$J_p \cong 0 = q\mu_p p_P \mathscr{E} - qD_p \frac{dp_P}{dx}$$

(3.11)

Thus the electric field is given* by

$$\mathscr{E} = \frac{kT}{q} \frac{1}{p_P} \frac{dp_P}{dx}$$

(3.12)

*Note that there is no electric field in a region of uniform majority carrier concentration.

The direction of this field is negative (since dp_P/dx is negative), aiding the motion of minority carriers (electrons) in this region. Thus the electron current is given by

$$J_n = q\mu_n n_P \mathscr{E} + qD_n \frac{dn_P}{dx} \tag{3.13a}$$

$$= qD_n \left(1 + \frac{n_P}{p_P}\right) \frac{dn_P}{dx} \tag{3.13b}$$

The limiting case for high level injection occurs when $n_P \cong p_P \to \infty$. For this case $J_n \to 2qD_n(dn_P/dx)$. Thus the effective diffusion constant for minority carriers doubles in the limit. Consequently,

$$J_n \cong -\frac{2qD_n \bar{n}_P}{L_n} e^{qV'_A/kT} \tag{3.14}$$

where V'_A is the voltage across the depletion layer.

The applied voltage across the diode is equal to the sum of the voltage across the depletion layer V'_A, and the voltage required to support the $\mathscr{E}$ field ($V_\mathscr{E}$). Hence

$$V'_A = V_A - V_\mathscr{E} \tag{3.15a}$$

$$= V_A + \int_0^W \mathscr{E}(x)\,dx \tag{3.15b}$$

For the limiting case of high level injection,

$$p_P(0) \cong n_P(0) = \bar{n}_P e^{qV'_A/kT} \tag{3.16}$$

In addition, $\bar{p}_P \bar{n}_P = n_i^2$. Solving (3.15b) and making these substitutions,

$$V'_A \cong V_A/2 + \frac{kT}{q} \ln \frac{n_i}{n_P} \tag{3.17}$$

Combining with (3.14) gives the forward current

$$J_{F,\text{high}} = J_n \cong -\frac{2qD_n n_i}{L_n} e^{qV_A/2kT} \tag{3.18}$$

Thus under high level conditions, the diffusion constant is effectively doubled and the voltage across the junction is halved. Furthermore, the doping level of the semiconductor becomes unimportant at high injection levels and is replaced by the intrinsic carrier concentration.

The transition voltage from low to high level injection is obtained by equating (3.9) and (3.18), it is given by

$$V_0 = \frac{2kT}{q} \ln \frac{2\bar{p}_P}{n_i} \tag{3.19}$$

For a diode where $\bar{p}_P = 10^{14}/cm^3$, this voltage is about 0.5 V. Power devices normally operate at forward voltages well in excess of this value, thus are commonly in high level injection.

3.1.1.1 The Narrow Base Diode

A diode is said to have a narrow base if the length of the lightly doped "base" region is less than the minority carrier diffusion length. This situation is encountered in some epitaxial devices of the punched-through type described in Section 2.4.2 and in the emitter-base diode of a transistor. In these structures the minority carrier concentration decays linearly with distance, not exponentially as in "long" structures, where the base width is greater than the minority carrier diffusion length. It can be shown that (3.9) and (3.18) apply to narrow base diodes also, provided the minority carrier diffusion length is replaced by the base width in both equations. Thus for low level injection in an n^+-p narrow base diode,

$$J_{F,\text{diff}} = -\frac{qD_{nB}\bar{n}_P}{W_B}(e^{qV_A/kT} - 1) \tag{3.20}$$

whereas for the high level case,

$$J_{F,\text{high}} \cong -\frac{2qD_{nB}n_i}{W_B}e^{qV_A/2kT} \tag{3.21}$$

Here D_{nB} is the minority carrier diffusion constant in the base region and W_B is the base width.

Associated with this diode is an effective transit time for minority carriers through the base region. If Q_B is the magnitude of excess base charge in a diode of cross-sectional area A, then

$$Q_B = \frac{qAW_B}{2}\left[n_P(0) - \bar{n}_P\right] \tag{3.22}$$

since $n_P(W_B) = \bar{n}_P$ at the ohmic contact. But

$$I_F = \frac{qAD_{nB}}{W_B} \left[n_P(0) - \bar{n}_P \right] \qquad (3.23)$$

From charge control considerations $Q_B = I_F t_B$, where t_B is the transit time. Thus for low level injection

$$t_B = \frac{W_B^2}{2D_{nB}} \qquad (3.24)$$

In like manner, it can be shown that the minority carrier transit time under high level injection conditions is given by

$$t_B \cong \frac{W_B^2}{4D_{nB}} \qquad (3.25)$$

since the effective diffusion constant is doubled for this situation.

3.1.2 Reverse Recovery

Consider an n^+-p diode in the circuit configuration of Fig. 3.2a. Initially the switch is in position A. Steady state conditions prevail at this point, and the magnitude of the forward diode current is given by $I_F \cong V_F/R$, if R is assumed to be large compared to the forward impedance. For this condition let the forward drop across the diode be V_D, where $V_D \ll V_F$.

At time $t = 0$ the switch is moved over to position B and a reverse voltage of magnitude V_{REV} is applied to the diode. This initiates the reverse recovery phase. Carriers moving by diffusion in the space charge neutral region are swept out of the diode in the reverse direction upon approaching the edge of the depletion layer. The velocity with which these carriers are moved through the depletion layer is on the order of the limiting velocity (about 10^7 cm/sec for electrons). Thus the current is limited only by the resistance in the external circuit. It has a magnitude of $I_R \cong V_{REV}/R$ and flows during the recovery phase. This reverse current is maintained as long as there is sufficient stored charge in the base.

At time $t = t_s$, this constant current phase is terminated. From here on the current to the external circuit falls off until it eventually reaches its reverse saturation value (which can be ignored for silicon).

Figure 3.3 illustrates the distribution of excess minority carriers in a wide base diode for $t \geqslant 0$. From $t = 0$ to $t = t_s$, the slope $(dn_P/dx)_{x=0}$ is constant, corresponding to the delivery of a constant reverse current to the

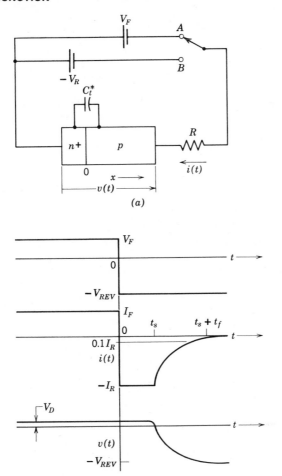

Fig. 3.2   Large signal characteristics.

circuit. From $t = t_s$ to infinity, this slope falls, as does the reverse current.

Waveforms of voltage and current are illustrated in Fig. 3.2b. Note that the diode remains forward biased until $t = t_s$. Beyond this point the voltage across it becomes negative, eventually reaching a magnitude of V_{REV}, since the reverse saturation current is ignored.

Consider an n^+-p diode of unit cross-sectional area. The continuity equation for electrons in the p-region may be written as

$$\frac{\partial \left(n_P - \bar{n}_P \right)}{\partial t} = \frac{n_P - \bar{n}_P}{\tau_n} - \frac{1}{q} \frac{\partial i_n(t)}{\partial x} \tag{3.26}$$

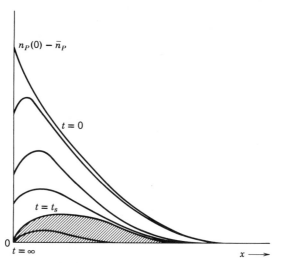

Fig. 3.3   Decay of stored charge.

where $i_n(t)$ is the current due to electron flow, defined positive in the direction shown in Fig. 3.2a.

The charge on the electron is $-q$ (where q is a positive number). Multiplying by $-q$ and integrating over the p region, we write

$$\frac{\partial}{\partial t} \int_0^W - q(n_P - \bar{n}_P)dx = \int_0^W \frac{\partial i_n(t)}{\partial x}dx - \int_0^W \frac{-q}{\tau_n}(n_P - \bar{n}_P)dx \quad (3.27)$$

Writing $Q(t)$ as the excess stored charge in the p-region at any given time, (3.27) reduces to

$$\frac{dQ(t)}{dt} + \frac{Q(t)}{\tau_n} - i_n(W,t) = -i_n(0,t) \quad (3.28)$$

The $Q(t)$ is associated with electrons in the base, thus is a negative quantity.

Define an effective lifetime in the forward direction, τ_F so that

$$\frac{Q(t)}{\tau_F} \equiv \frac{Q(t)}{\tau_n} - i_n(W,t) \quad (3.29)$$

Substitution of (3.29) into (3.28) gives

$$\frac{dQ(t)}{dt} + \frac{Q(t)}{\tau_F} = -i_n(0,t) \tag{3.30}$$

The total forward current of an n^+-p diode consists of the current through the depletion layer capacitance as well as of the current due to electron flow. If we write the average depletion layer capacitance as C_t^*, (3.30) may be modified to

$$\frac{dQ(t)}{dt} + \frac{Q(t)}{\tau_F} - C_t^* \frac{dv(t)}{dt} = -i(t) \tag{3.31}$$

where C_t^* is the average capacitance value and is given by

$$C_t^* = \frac{1}{V_2 - V_1} \int_{V_1}^{V_2} C_t(v) \, dv \tag{3.32}$$

Here V_1 and V_2 are the limits over which the voltage across the diode varies.

Equation 3.31 is a statement of *the charge control principle*. Since it does not involve the detailed behavior of minority carriers in the semiconductor, it can be used with different impurity configurations.

Before reversal of the voltage, that is, at $t \leq 0, i(t) = I_F$. Thus the steady state excess stored charge is $-I_F \tau_F$. Upon reversal of the voltage, a constant current $-I_R$ flows during the time interval from 0 to t_s. During this interval the change in the voltage drop across the diode is small (see Fig. 3.2b). Consequently the current through the depletion layer capacitance may be ignored. Hence (3.31) reduces to

$$\frac{dQ(t)}{dt} + \frac{Q(t)}{\tau_F} = I_R \tag{3.33}$$

Solving this equation, subject to the initial value condition, gives

$$Q(t) = I_R \tau_F - (I_R + I_F) \tau_F e^{-t/\tau_F} \tag{3.34}$$

It is necessary to make some assumptions concerning the charge remaining in the p-region at $t > t_s$. Let us assume that the carrier concentration can be written as

$$n(x,t) = f(t) n(x) \tag{3.35a}$$

where $f(t)$ is a decaying function of time. Then, it follows that the stored charge at any given time is directly proportional to the reverse current. We can thus write

$$Q(t) = -i(t)\tau_R \tag{3.35b}$$

where τ_R is an effective reverse lifetime. Then the excess stored charge at $t = t_s$ is given by $-I_R\tau_R$. Substituting this condition into (3.34) gives

$$t_s = \tau_F\left[\ln\left(1 + \frac{I_F}{I_R}\right) - \ln\left(1 + \frac{\tau_R}{\tau_F}\right)\right] \tag{3.36}$$

Measurements of t_s for two different values of I_F/I_R are necessary to determine both τ_F and τ_R. These parameters can then be used to characterize a diode, regardless of its doping configuration or structure.

For a wide base diode, $i_n(W,t) = 0$. Thus τ_F is equal to the minority carrier lifetime. For the narrow base diode, however,

$$i_n(W,t) \cong \frac{Q(t)}{t_B} \tag{3.37}$$

where t_B is the transit time of carriers through the base. Thus a contact close to the depletion layer acts as a sink for diffusing electrons and reduces the effective lifetime in the narrow base region. Combining with (3.29),

$$1/\tau_F = 1/\tau_n + 1/t_B \tag{3.38}$$

The ratio τ_F/τ_R is equal to the ratio of the excess charge stored in the base under forward bias to the excess charge stored at $t = t_s$. This ratio is a sensitive function of the doping profile and must be experimentally determined. It is usually found to lie between 2 and 4.

The current through C_t^* cannot be ignored during the fall period ($t \geqslant t_s$). From the loop equation for the network of Fig. 3.2a it is seen that $dv(t) = -R\,di(t)$. Substituting this relationship and (3.35b) into (3.31), the equation determining the fall period is

$$\frac{di(t)}{dt}(\tau_R + RC_t^*) + i(t)\left(1 + \frac{\tau_R}{\tau_F}\right) = 0 \tag{3.39}$$

At $t = t_s$, $i(t) = -I_R$. Thus

$$i(t) = -I_R\exp\left[\frac{-(t - t_s)}{\tau}\right] \tag{3.40}$$

where

$$\tau = \frac{\tau_R + RC_t^*}{1 + \tau_R/\tau_F} \tag{3.41}$$

The fall time, measured to 10% of the initial value of reverse current, is written as

$$t_f \cong \frac{2.3(\tau_R + RC_1^*)}{1 + \tau_R/\tau_F} \tag{3.42}$$

It is thus seen that the recovery characteristics of a diode are related to the minority carrier lifetime, to the transit time (for a narrow base diode), to the forward and reverse currents, and to circuit parameters. In addition, the manner in which the device is fabricated plays a role in determining τ_R.

3.2 THE DIFFUSED JUNCTION

Diffused junctions are commonly encountered in high voltage rectifiers, as well as in power transistors. The electric field in a diffused n^+-p junction has been shown to be oriented in the $n^+ \rightarrow p$ direction (see Section 2.4.5). Furthermore, its magnitude in the n^+-region is considerably larger than in the p-region because of the exponential nature of the impurity grading. As a consequence, forward current is carried by minority carrier injection of electrons into a weakly retarding field, and by holes into a strongly retarding field. Thus both these components of diode current are correspondingly altered.

The following analysis [3] applies to arbitrarily doped junctions, in which N is the net ionized impurity concentration at any point (donors taken as positive), and n and p are the electron and hole concentration, respectively. The origin is taken at the junction, with the n-region on the left side and the p-region on the right. Finally, the depletion layer width is ignored. Note that N, n, p, D_n, and D_p are all functions of distance.

Under steady state, forward bias conditions, the current density due to electron flow in the p-region is given by

$$J_n = q\mu_n n_P \mathscr{E} + qD_n \frac{dn_P}{dx} \tag{3.43}$$

The electric field in the p-region is given by

$$\mathscr{E} = \frac{kT}{q} \frac{1}{N} \frac{dN}{dx} \tag{3.44}$$

Combining these equations with the Einstein relation gives

$$J_n = \frac{qD_n}{N}\left(N\frac{dn_P}{dx} + n_P\frac{dN}{dx} \right) \tag{3.45}$$

$$= \frac{qD_n}{N}\frac{d(Nn_P)}{dx} \tag{3.46}$$

Let W_P be the base width of the p-region, so that $n_P = \bar{n}_P$ at $x = W_P$. Assume that W_P is small compared to the diffusion length of minority carriers, so that J_n is constant over this region. For this condition, (3.46) can be integrated to give

$$N(n_P - \bar{n}_P) = -\frac{J_n}{q}\int_x^{W_P}\frac{N}{D_n}dx \tag{3.47}$$

At the edge of the depletion layer,

$$N(0)\left[n_P(0) - \bar{n}_P \right] = -\frac{J_n}{q}\int_0^{W_P}\frac{N}{D_n}dx \tag{3.48}$$

But

$$n_P(0) = \bar{n}_P(0)e^{qV_A/kT} \tag{3.49}$$

where $\bar{n}_P(0)$ is the equilibrium concentration at the edge of the depletion layer in the p-region and is equal to $n_i^2/N(0)$. Combining these equations,

$$J_n = -\frac{qn_i^2}{\int_0^{W_P}(N/D_n)dx}(e^{qV_A/kT} - 1) \tag{3.50}$$

In like manner, the current density due to hole flow is

$$J_p = -\frac{qn_i^2}{\int_{W_N}^0(N/D_p)dx}(e^{qV_A/kT} - 1) \tag{3.51}$$

where the width of the n-region W_N is short compared to the hole diffusion length. The total current density is thus

$$J = -qn_i^2 \left[\frac{D_p^*}{\int_{W_N}^0 N\,dx} + \frac{D_n^*}{\int_0^{W_P} N\,dx} \right] (e^{qV_A/kT} - 1) \qquad (3.52)$$

where D_p^* and D_n^* are average values of the diffusion constants.

This relation is of the same general form as that for the abrupt diode. If either region is long, it must be modified to include the effect of minority carrier diffusion. For example, if the n^+-region is long, the injected hole current density falls off exponentially with distance and is given* by $J_p(0)e^{-x/L_p}$. Making this substitution results in

$$J_p \cong -\frac{qn_i^2}{\int_{W_N}^0 (N/D_p)e^{-x/L_p}\,dx} (e^{qV_A/kT} - 1) \qquad (3.53)$$

The total forward current density of an n^+-p diode, having a long n^+-region and a short p-region, is thus given by

$$J \cong -qn_i^2 \left[\frac{D_p^*}{\int_{W_N}^0 Ne^{-x/L_p}\,dx} + \frac{D_n^*}{\int_0^{W_P} N\,dx} \right] (e^{qV_A/kT} - 1) \qquad (3.54)$$

It is important to keep the ratio of electron to hole current density high in n-p-n transistors, to achieve a high injection efficiency (see Section 4.3.1). This is done by doping the emitter region heavily.

3.2.1 Heavy Doping Effects

One effect of heavy doping of the emitter is the falling of the fraction of ionized impurities in this region as the Fermi level moves toward the band edge (and eventually into the conduction band for an n^+-emitter). This problem has been dealt with elsewhere [2] and is not repeated here. A second effect is the reduction of the energy gap. For phosphorus-doped silicon, this reduction is significant at doping levels above $1.85 \times 10^{19}/cm^3$

---

*This is only true if W_N is infinitely long. Thus (3.53) is a compromise if W_N is of finite length.

and is given by [4]

$$\Delta E_g = 3.4 \times 10^{-8} \left(N_d^{1/3} - 2.65 \times 10^6 \right) \text{eV} \tag{3.55}$$

The reduction in energy gap occurs over the width of the heavily doped emitter region and lowers the retarding electric field here. Under extreme conditions, it has been shown [5] that the electric field can even reverse its direction over part of the emitter region, thereby aiding the hole flow. Thus the hole current injected into the emitter increases with heavy doping. Note that the electron current injected into the p-type base region is unchanged because of its relatively light doping. As a consequence, heavy doping of the emitter region lowers the ratio of injected electron to hole current. This results in a fall in the injection efficiency for an n^+-p-n transistor.

Consider a forward-biased diffused n^+-p diode [6]. Let N be the net ionized impurity concentration at any point, as before. Note that N, n, p, D_n, D_p, and n_i are all functions of distance. The metallurgical junction is taken as the origin. The injected electron current density in the n^+-region is given by

$$J_n = q\mu_n n \mathcal{E} + qD_n \frac{dn}{dx} \tag{3.56}$$

In the absence of energy gap narrowing, the electric field is obtained by setting the equilibrium electron current density to zero. Combining (3.56) with the Einstein relation, the field in the n^+-region is given by

$$\mathcal{E} = -\frac{kT}{q} \frac{1}{N} \frac{dN}{dx} \tag{3.57}$$

The effect of energy gap narrowing is to alter this electric field, so that

$$\mathcal{E} = -\left[\frac{1}{q} \frac{d}{dx} (\Delta E_g) + \frac{kT}{q} \frac{1}{N} \frac{dN}{dx} \right] \tag{3.58}$$

where ΔE_g is the reduction in the energy gap.

The hole current density is given by substituting (3.58) into (3.56), to yield

$$-\frac{J_p}{qD_p} = \left[\frac{1}{kT} \frac{d}{dx} (\Delta E_g) + \frac{1}{N} \frac{dN}{dx} \right] + \frac{dp}{dx} \tag{3.59}$$

This equation can be directly solved if J_p is constant over the n-region, that is, if W_N is short compared to the minority carrier diffusion length. For this situation, the boundary values are

$$p_N(-W_N) = \bar{p}_N \qquad (3.60a)$$

and

$$p_N(0) = \bar{p}_N(0) e^{qV_A/kT}$$

$$= \frac{n_{i0}^2}{N(0)} e^{qV_A/kT} \qquad (3.60b)$$

where n_{i0} is the intrinsic carrier concentration in the absence of energy gap narrowing. Solving,

$$J_p \cong - \left[\frac{q n_{i0}^2}{\int_{W_N}^0 (N/D_p) e^{-\Delta E_g/kT} dx} \right] (e^{qV_A/kT} - 1) \qquad (3.61a)$$

$$\cong - \left[\frac{q D_p^* n_{i0}^2}{\int_{W_N}^0 N e^{-\Delta E_g/kT} dx} \right] (e^{qV_A/kT} - 1) \qquad (3.61b)$$

where D_p^* is the average value of D_p.

A similar solution for J_n can be obtained if the p-region is short compared to the diffusion length of injected electrons. In this case, however, $\Delta E_g = 0$ because of the relatively light doping level. Thus

$$J_n \cong - \left[\frac{q D_n^* n_{i0}^2}{\int_0^{W_P} N dx} \right] (e^{qV_A/kT} - 1) \qquad (3.62)$$

The total current density is given by the sum of these separate components. Once again, these equations must be modified if long regions are involved, to take into account the minority carrier diffusion length. For example, for a long n^+-region, the decay in hole current density with distance can be approximated by e^{-x/L_p}, so that

$$J_p \cong - \left[\frac{qD_p^* n_{i0}^2}{\int_{W_N}^0 N e^{-\Delta E_g/kT} e^{-x/L_p} dx} \right] (e^{qV_A/kT} - 1) \qquad (3.63)$$

Diffusion lengths in heavily doped n^+-regions range typically from as low as 0.15 μm in alloyed regions up to 12 μm in diffused regions.

3.3   THE p^+-i-n^+ DIODE

The p^+-i-n^+ diode is especially important because it is the most effective structure for supporting a high reverse voltage. Furthermore, the forward characteristics of a thyristor in its ON state closely resemble those of a p^+-i-n^+ device, as Chapter 5 demonstrates. This section considers in detail the characteristics of this type of device.

Suppose that in an abrupt junction p^+-i-n^+ diode, $2d$ is the base width of the i-region, and this width is on the order of the diffusion length for either electrons or holes. Under forward bias conditions, $n = p \gg n_i$. As a result, high level injection conditions prevail in the mid-region of this device.

For the present, assume that the injection efficiency at both ends is unity. Thus the components of forward current flow due to minority carriers in the p^+- and n^+-regions can be neglected, and current flow is accounted for by the rate at which holes and electrons recombine within the i-region. The current density is thus given by

$$J = \int_{-d}^d qR \, dx \qquad (3.64)$$

where R is the recombination rate and is equal to $n(x)/\tau$. If n^* is the average injected electron concentration in the i-region, then

$$J = \frac{2qn^*d}{\tau_a} \qquad (3.65)$$

where the ambipolar lifetime τ_a is equal to the high level lifetime $\tau_{p0} + \tau_{n0}$. Further simplification can be made if the carrier concentration is considered to be approximately constant throughout the i-region, permitting us to neglect carrier motion by diffusion. Since $n^* \cong p^*$,

$$J = q(\mu_n + \mu_p)n^* \mathcal{E}^* \qquad (3.66a)$$

$$= \frac{q}{kT}\left(1 + \frac{1}{b}\right)qD_n n^* \mathcal{E}^* \qquad (3.66b)$$

where $b = \mu_n / \mu_p$, and $\mathcal{E}^*$ is the average electric field. Combining with (1.27) for the high level ambipolar diffusion constant gives

$$J = \frac{q}{kT}\frac{(b+1)^2}{2b}qD_a n^* \mathcal{E}^* \qquad (3.67)$$

Finally, the voltage drop across the mid-region V_M is given by

$$V_M = 2d\mathcal{E}^* \qquad (3.68)$$

Combining (3.66b), (3.67), and (3.68), and noting that $L_a = (D_a \tau_a)^{1/2}$,

$$V_M = \frac{kT}{q}\frac{8b}{(1+b)^2}\left(\frac{d}{L_a}\right)^2 \qquad (3.69a)$$

For the equal mobility case, $b = 1$ and

$$V_M = \frac{2kT}{q}\left(\frac{d}{L_a}\right)^2 \qquad (3.69b)$$

For silicon, where $b = 3$,

$$V_M = \frac{3kT}{2q}\left(\frac{d}{L_a}\right)^2 \qquad (3.69c)$$

The voltage drop across the i-region is thus independent of the current through it. This is a direct consequence of the assumption that all the forward current is due to an increase in the electron and hole concentrations over their equilibrium values, with a concurrent increase in the conductivity of the i-region.

For example, if $d = L_a = 200$ μm, the average electric field is 0.97 V/cm, and the voltage drop across the i-region is 39 mV. This serves to further emphasize the need for a long lifetime process, since $L_a = (D_a \tau_a)^{1/2}$.

Let us now set forth a more complete theory [7, 8] for the p^+-i-n^+ diode. Figure 3.4 shows this device with the central region assumed to be ν-type

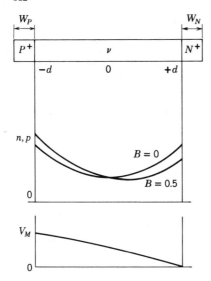

Fig. 3.4   The p^+-ν-n^+ **diode.**

for the sake of specificity. The length of the ν-region is taken as $2d$. As before, unit injection efficiency is assumed for the two junctions.

The steady state equations for carrier transport in high resistivity materials can be written (see Section 1.2) as

$$\frac{\partial n}{\partial t} = 0 = -\frac{n}{\tau_a} + D_a \frac{\partial^2 n}{\partial x^2} \tag{3.70}$$

since $n' = n \gg N_\nu$. But $L_a = (D_a \tau_a)^{1/2}$, so that

$$\frac{d^2 n}{dx^2} - \frac{n}{L_a^2} = 0 \tag{3.71}$$

Here D_a and L_a are the ambipolar parameters, and τ_a is the high level lifetime. The boundary values are next determined.

The current density is given by

$$J = J_{n, +d} = q\mu_n n_{+d} \mathcal{E}_{+d} + q D_n \frac{dn}{dx}\bigg|_{x=+d} \tag{3.72}$$

But

$$\mathcal{E}_{+d} = \frac{kT}{q} \frac{1}{n_{+d}} \frac{dn}{dx}\bigg|_{x=+d} \tag{3.73}$$

so that

$$J = 2qD_n \frac{dn}{dx}\bigg|_{x=+d} \tag{3.74a}$$

Similarly,

$$J = J_{p,-d} = -2qD_p \frac{dn}{dx}\bigg|_{x=-d} \tag{3.74b}$$

Solving (3.70), and combining with (3.73) and (3.74), gives the electron concentration as

$$n = \frac{\tau_a J}{2qL_a}\left[\frac{\cosh(x/L_a)}{\sinh(d/L_a)} - B\frac{\sinh(x/L_a)}{\cosh(d/L_a)}\right] \tag{3.75}$$

where

$$B = \frac{\mu_n - \mu_p}{\mu_n + \mu_p} \tag{3.76}$$

For silicon, $b = \mu_n/\mu_p = 3$, so that $B = 0.5$. If the electron and hole mobilities are assumed equal, then $B = 0$. Figure 3.4 shows this function for these two cases. Note the asymmetry that arises because of the unequal values of electron and hole mobility.

Now we can determine the voltage drop across the ν-region:

$$J_p = q\mu_p\left(n\mathscr{E} - \frac{kT}{q}\frac{dn}{dx}\right) \tag{3.77}$$

$$J_n = q\mu_n\left(n\mathscr{E} + \frac{kT}{q}\frac{dn}{dx}\right) \tag{3.78}$$

and

$$J = J_p + J_n \tag{3.79}$$

Combining these equations gives

$$\mathscr{E} = \frac{J}{q(\mu_n + \mu_p)n} - \frac{kT}{q}\frac{B}{n}\frac{dn}{dx} \tag{3.80}$$

The first of these terms is due to an ohmic drop, whereas the second is caused by the asymmetric concentration gradient produced by the unequal mobilities. The voltage drop across the mid-region is obtained by integrating (3.80) over its length, resulting in

$$
\frac{V_M}{kT/q} = \left\{ \frac{8b}{(b+1)^2} \frac{\sinh(d/L_a)}{\sqrt{1 - B^2\tanh^2(d/L_a)}} \right.
$$

$$
\cdot \arctan\left[\sqrt{1 - B^2\tanh^2\left(\frac{d}{L_a}\right)} \ \sinh\left(\frac{d}{L_a}\right)\right]\right\}
$$

$$
+ B\ln\left[\frac{1 + B\tanh^2(d/L_a)}{1 - B\tanh^2(d/L_a)} \right] \tag{3.81}
$$

Figure 3.5 plots this equation. For values of $d \leqslant L_a$, it can be approximated

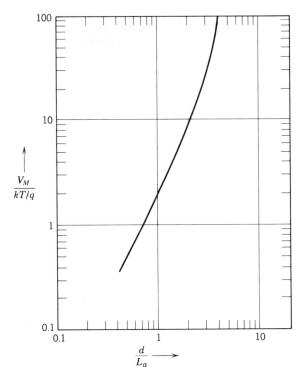

Fig. 3.5 V_M as a function of d/L_a.

[8] by

$$V_M \cong \frac{3kT}{2q} \left(\frac{d}{L_a} \right)^2 \tag{3.82a}$$

and by

$$V_M \cong \frac{3\pi kT}{8q} \exp\left(\frac{d}{L_a} \right) \tag{3.82b}$$

for $d \geqslant 3L_a$.

Equations 3.75, 3.80, and 3.81 can be considerably simplified for the equal mobility case, where $b = 1$ and $B = 0$. For this situation,

$$n = \frac{\tau_a J}{2q L_a} \frac{\cosh(x/L_a)}{\sinh(d/L_a)} \tag{3.83}$$

$$\mathscr{E} = \frac{kT}{q L_a} \frac{\sinh(d/L_a)}{\cosh(x/L_a)} \tag{3.84}$$

and the voltage drop across the mid-region is then

$$V_M = \frac{2kT}{q} \left[\sinh\left(\frac{d}{L_a} \right) \right] \left\{ \text{arc tan} \left[\sinh\left(\frac{d}{L_a} \right) \right] \right\} \tag{3.85}$$

For values of $d \leqslant L_a$, (3.85) reduces to

$$V_M \cong \frac{2kT}{q} \left(\frac{d}{L_a} \right)^2 \tag{3.86}$$

The condition $d = L_a$ thus marks the transition between a "short" and a "long" p^+-i-n^+ diode. For the short structure, where $d \leqslant L_a$, the spatial variation in carrier concentration over the i-region has little effect on the voltage drop and can be ignored. In high voltage structures, however, values of d can be as large as 200–300 μm (base widths of 400–600 μm), and device structures are usually long. Now, the effect of the spatial variation of carrier concentration becomes decisive in establishing device characteristics.

Finally, it will be shown that the diffusion length falls with injected carrier concentration; therefore a device that behaves like a short structure at low forward current densities can effectively become a long structure at high levels. Thus device performance can become seriously impaired at surge current densities, which often exceed $1000 \text{ A}/\text{cm}^2$.

3.3.1 The Forward Voltage

The total applied voltage is given by

$$V_A = V_L + V_M + V_R \tag{3.87}$$

where V_L and V_R are the voltage drops across the left and right junctions of the device shown in Fig. 3.4. The electron and hole concentrations at the two ends of the mid-region are related to their equilibrium values, such that

$$\frac{p_{-d}}{n_i^2 / N_\nu^+} = e^{qV_L/kT} \tag{3.88a}$$

$$\frac{n_{+d}}{N_\nu^+} = e^{qV_R/kT} \tag{3.88b}$$

Since $p_{-d} = n_{-d}$, this gives

$$V_L + V_R = \frac{kT}{q}\left(\ln \frac{n_{+d} n_{-d}}{n_i^2} \right) \tag{3.89}$$

Here, N_ν^+ is the ionized donor concentration in the ν-region, and n_{+d}, n_{-d} are the electron concentrations in the mid-region at $-d, +d$, respectively, as obtained from (3.75). Typically, the sum of these two voltages is approximately equal to that for two forward-biased diodes. The equation for the diode may be written in closed form by combining (3.75) with (3.87) to (3.89), resulting in

$$J = \frac{2qD_a n_i}{d} F\left(\frac{d}{L_a} \right) e^{qV_A/2kT} \tag{3.90}$$

where

$$F\left(\frac{d}{L_a} \right) = \left[\frac{d}{L_a}\tanh\left(\frac{d}{L_a} \right) \right]\left[1 - B^2\tanh^4\left(\frac{d}{L_a} \right) \right]^{-1/2} e^{-qV_M/2kT} \tag{3.91}$$

Equation 3.90 bears a striking similarity to that for high level injection in an n^+-p diode. Note that $F(d/L_a)$ is independent of current density in this equation.

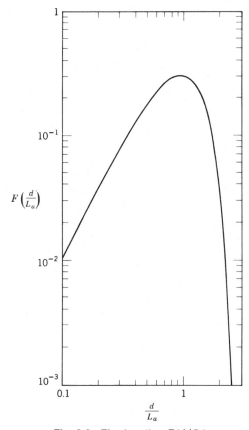

Fig. 3.6   The function $F(d/L_a)$.

Figure 3.6 presents $F(d/L_a)$ as a function of d/L_a. Since this function peaks at $d/L_a = 1$, it would appear that the forward voltage drop is minimized at this point. In practice, however, it has been found [9] that the forward voltage falls rapidly with increasing lifetime, until $d/L_a \cong 1$. For still larger values of lifetime, this voltage drop remains constant or falls only very slightly. This is because a number of additional factors (e.g., carrier-carrier scattering, Auger recombination, and injection efficiency effects) must be considered before obtaining a true assessment of the forward voltage.

Carrier-carrier scattering serves to reduce the ambipolar diffusion constant at high injection levels, whereas Auger recombination reduces the effective lifetime. Together, these act to reduce the diffusion length, hence increase the d/L_a ratio* at high injection levels. Their effect is particularly significant when the carrier concentration exceeds $10^{17}/\text{cm}^3$.

The effect of recombination in the p^+- and n^+-regions is twofold. To a small extent, it necessitates higher values of V_L and V_R to maintain the appropriate carrier concentrations in the ν-region. More important, however, it reduces the injection efficiency, hence the fraction of the forward current that is available for conductivity modulation of the base region. Typically, for devices with mid-region lifetimes in excess of 10 μsec, it has been shown [10] that about half the forward current is due to recombination in the end regions at current densities below 5 A/cm². However this number exceeds 90% at current densities above 1000 A/cm². Thus less than 10% of the forward current contributes to conductivity modulation of the ν-region at surge current levels, and the forward voltage drop is greatly increased.

The detailed nature of these three effects, and their influence on the device behavior, is now considered.

3.3.1.1 Carrier-Carrier Scattering Effects

Mobility in semiconductors is affected by a number of scattering processes, and the most common are lattice and ionized impurity scatter.

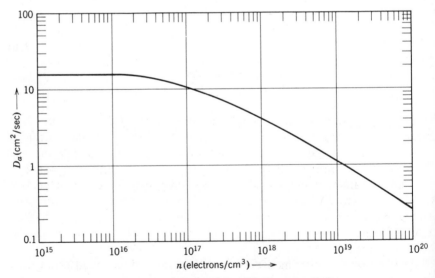

Fig. 3.7   The ambipolar diffusion constant.

---

*Note that d is set by the reverse voltage requirement on the device and is a design parameter.

Carrier-carrier scattering is the process by which mobile carriers interact. It is similar to ionized impurity scattering, except that mobile particles deflect about a common center of mass. Carrier-carrier scattering between particles of the same charge type results in no net effect on the conductivity and can be ignored. On the other hand, the unequal mass of electrons and holes ($m_n = 0.26\,m_0$, $m_p = 0.38\,m_0$, where m_0 is the mass of an electron in free space) can lead to significant scattering effects in the ν-region of a p^+-i-n^+ diode at high injected carrier concentrations.

The arguments for ionized impurity scatter, which have been extended to include the effects of carrier-carrier scatter [11], result in an ambipolar diffusion constant whose value falls with the injected carrier concentrations (Fig. 3.7). Note that the reduction in D_a is particularly significant at injected carrier concentrations in excess of $10^{17}/\mathrm{cm}^3$. For $d = 200$ μm and $\tau_a = 10$ μsec, this corresponds to current densities in excess of 64 A/cm^2.

3.3.1.2 Auger Recombination Effects

Section 1.1 demonstrated that Auger recombination processes dominate the behavior of minority carrier lifetime at high injection levels. Here the effective lifetime can be written as

$$\frac{1}{\tau_{\mathrm{eff}}} = \frac{1}{\tau_a} + \gamma_3 n^2 \tag{3.92}$$

where τ_a is the ambipolar lifetime ($= \tau_{p0} + \tau_{n0}$) and γ_3 ranges from 2 to 2.9×10^{-31} $\mathrm{cm}^6/\mathrm{sec}$ for silicon. Calculated values of τ_{eff}, based on $\gamma_3 = 2.9 \times 10^{-31}$ $\mathrm{cm}^6/\mathrm{sec}$, appear in Fig. 3.8 for a number of values of τ_a, and have been verified experimentally [12]. Again, it is interesting to note that τ_{eff} falls at injected carrier concentration levels in excess of $10^{17}/\mathrm{cm}^3$.

In addition to altering the d/L_a ratio, Auger processes play an important role in determining the variation of the injected carrier concentration with the forward current density. Thus for $n \leqslant 10^{17}/\mathrm{cm}^3$,

$$J = \frac{2qdn^*}{\tau_a} \tag{3.93a}$$

so that $J \propto n^*$. Once Auger processes become important however (at $n > 10^{17}/\mathrm{cm}^3$), this relationship changes. In the limit,

$$J = \frac{2qd(n^*)^3}{\gamma_3} \tag{3.93b}$$

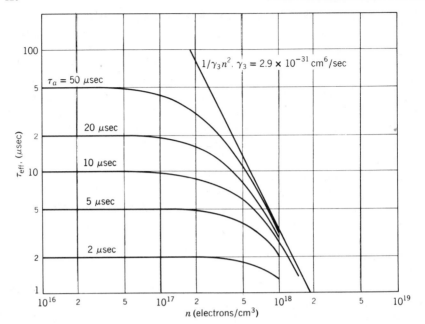

Fig. 3.8   The effective lifetime.

so that $J \propto (n^*)^3$. Thus the variation of forward current density with injected carrier concentration is greatly increased.

3.3.1.3   Injection Efficiency Effects

The analytical treatment so far has been based on the assumption of unit injection efficiency for both the p^+-ν and the ν-n^+ junctions. Thus all the forward current was assumed to result in minority carrier injection into the ν-region. In practice, however, this forward current is also due to the injection of electrons into the p^+-region and holes into the n^+-region. These components, which do not contribute to conductivity modulation of the ν-region, greatly increase the voltage drop across the diode [13, 14].

Doping levels in the p^+- and n^+-regions typically exceed $10^{19}/\text{cm}^3$. Thus the injected electron current density in the end regions can be obtained from low level injection theory, and is given by (3.10) as

$$J_{nP} = \left[\frac{q D_{nP} \bar{n}_P}{L_{nP} \tanh(W_P / L_{nP})} \right] e^{q V_L / kT} \qquad (3.94a)$$

Where $\bar{n}_P$, D_{nP}, L_{nP}, and W_P refer to the p^+-region. In like manner,

$$J_{pN} = \left[\frac{q D_{pN} \bar{p}_N}{L_{pN} \tanh(W_N / L_{pN})} \right] e^{qV_R/kT} \tag{3.94b}$$

where $\bar{p}_N$, D_{pN}, L_{pN}, and W_N refer to the n^+-region. These equations can be written more compactly as

$$J_{nP} = J_{ns} e^{qV_L/kT} \tag{3.95a}$$

$$J_{pN} = J_{ps} e^{qV_R/kT} \tag{3.95b}$$

where J_{ns} and J_{ps} are the saturation current densities of the heavily doped regions. In practical devices, heavily doped p^+-regions are produced by alloying with aluminum, or by the diffusion of gallium, aluminum, and boron, or their combinations (see Chapter 6). Heavily doped n^+-regions are commonly made by alloying with antimony-doped gold-germanium eutectic preforms or by phosphorus diffusion. Thus J_{ns} and J_{ps} are usually not equal in magnitude. Typical values range from 10^{-13} to 4×10^{-13} A/cm^2 for a wide variety of structures, with J_{ns} usually less than J_{ps} by a factor of 1.5 to 2.

Under quasi–thermal equilibrium conditions, the electron and hole concentrations on either side of a depletion layer are interrelated such that

$$\frac{p_{P,-d}}{p_{-d}} = \frac{n_{-d}}{n_{P,-d}} \tag{3.96}$$

where $p_{P,-d}, n_{P,-d}$ refer to the depletion layer edge on the heavily doped p^+ side. But $p_{P,-d} \cong \bar{p}_P$, and $n_{P,-d} = \bar{n}_P e^{qV_L/kT}$, so that

$$n_{-d} p_{-d} \cong (n_{-d})^2 = \bar{p}_P \bar{n}_P e^{qV_L/kT} \tag{3.97a}$$

$$= n_i^2 e^{qV_L/kT} \tag{3.97b}$$

Combining with (3.95a),

$$J_{nP} = J_{ns} \left(\frac{n_{-d}}{n_i} \right)^2 \tag{3.98a}$$

Similarly,

$$J_{pN} = J_{ps} \left(\frac{n_{+d}}{n_i} \right)^2 \tag{3.98b}$$

It should be noted that the forward current due to recombination in the end regions varies as $(n^*)^2$, since n^* is linearly related to n_{+d} and n_{-d}. In contrast, the forward current due to recombination in the mid-region varies linearly with n^* at low levels, as shown by (3.93a). With growing carrier concentration, the recombination current in the end regions be- comes an increasingly significant fraction of the total current. At surge current levels, more than 90% of the forward current is supplied by this recombination current.

The total forward current density J_F is given by

$$J_F = J_{nP} + J + J_{pN} \tag{3.99}$$

where J is the current density due to minority carrier injection into the ν-region. The fraction of the forward current that contributes to conductiv- ity modulation of the ν-region is J/J_F. To an approximation, therefore, the voltage drop across the mid-region of a p^+-i-n^+ diode (in which recombi- nation in the end regions is significant) is given by V'_M, where

$$V'_M \cong \frac{J_F}{J} V_M \tag{3.100}$$

and V_M is given by (3.81).

Detailed calculations [14] show that the actual value of V_M is now a function of a symmetry factor $s = (bJ_{ps}/J_{ns})^{1/2}$, but is only slightly changed from the value given by (3.81) for the case when recombination in the end regions is ignored. This is illustrated in Fig. 3.9, where $V_M/(kT/q)$ is plotted for different values of s, as well as for the unit injection efficiency case ($J_{ns} = J_{ps} = 0$).

The forward characteristic can be readily calculated for the equal mobility case, where $B = 0$. From (3.83),

$$n_{+d} = n_{-d} = \frac{J\tau_a \coth(d/L_a)}{2qL_a} \tag{3.101}$$

so that

$$J_{nP} = J_{ns} \left(\frac{\tau_a \coth(d/L_a)}{2qn_i L_a} \right)^2 J^2 \tag{3.102a}$$

and

$$J_{pN} = J_{ps} \left(\frac{\tau_a \coth(d/L_a)}{2qn_i L_a} \right)^2 J^2 \tag{3.102b}$$

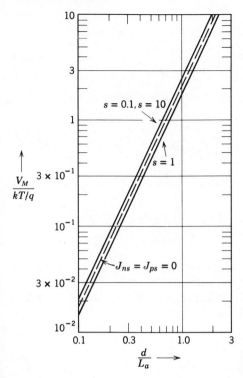

Fig. 3.9 V_M as a function of s and d/L_a. Copyright © 1968 by Solid State Electronics. Reprinted with permission from Herlet [14].

For convenience, these equations may be written as

$$J_{nP} = K_1 J^2 \qquad (3.103a)$$

$$J_{pN} = K_2 J^2 \qquad (3.103b)$$

where K_1 and K_2 are constant for any specific device. Combining with (3.99) for the forward current density results in

$$J_F = \frac{1}{K_1 + K_2} \frac{J_F}{J}\left(\frac{J_F}{J} - 1\right) \qquad (3.104)$$

Detailed calculations for $b \neq 0$ [14] for the forward current density show that

$$J_F = \frac{\sqrt{b}}{b+1} \frac{(s+1)^2}{s} \frac{(qn_i D_a/L_a)^2}{(J_{ps}J_{ns})^{1/2}} \tanh^2\left(\frac{d}{L_a}\right) \frac{J_F}{J}\left(\frac{J_F}{J} - 1\right) \qquad (3.105)$$

This equation is of the same form as (3.104), that is,

$$J_F = K \frac{J_F}{J} \left(\frac{J_F}{J} - 1 \right) \tag{3.106}$$

where K is a constant for any specific device. Combining with (3.99)

$$J_F = K \frac{V'_M}{V_M} \left(\frac{V'_M}{V_M} - 1 \right) \tag{3.107}$$

The voltage drop across the end regions is given for the zero current case by the contact potential, that is,

$$V_L + V_R + V_D = \left(\frac{kT}{q} \right) \ln \left(\frac{p_P n_N}{n_i^2} \right) \tag{3.108}$$

where p_P and n_N are the majority carrier concentrations in the end regions. To a good approximation, this term varies only slightly with increasing forward current, since the low level injection condition is met for the end regions. Thus (3.107) can be rewritten

$$J_F = K \frac{V_F - V_D}{V_M} \left(\frac{V_F - V_D}{V_M} - 1 \right) \tag{3.109}$$

where V_F is the forward voltage drop across the p^+-i-n^+ diode. Note that at high levels, this device exhibits a square law characteristic.

In summary, it has been shown that the forward current density of a p^+-i-n^+ diode is given by

$$J_F = J_{nP} + J + J_{pN} \tag{3.110}$$

where J is due to recombination in the mid-region, and J_{nP}, J_{pN} are due to recombination in the p^+- and n^+-regions, respectively. These terms are related to the average injected carrier concentration in the mid-region (n^*) as follows:

$$
\begin{aligned}
J &\propto n^* & &\text{for} \quad n^* < 10^{17} \\
J &\propto (n^*)^3 & &\text{for} \quad n^* \gg 10^{17} \\
J_{nP} &\propto (n^*)^2 & & \\
J_{pN} &\propto (n^*)^2 & &
\end{aligned}
$$

At low levels, the contributions of J_{nP} and J_{pN} are small relative to J, and Auger effects can be ignored. Thus the variation of n^* with J is essentially linear. At high levels, however, a greater fraction of the total current is carried by these terms until the average injected carrier concentration varies as the square root of the forward current density. With the onset of Auger recombination, the variation of n^* becomes more sublinear with J and eventually takes on a cube root dependence.

It has been shown that a number of factors cooperate in determining the forward characteristics of the p^+-i-n^+ diode. Their various effects are as follows:

1. A diode may be either "long" or "short" depending on whether d/L_a is greater or less than unity. The voltage drop across the mid-region of a long diode is considerably larger than that across a short diode.
2. Both carrier-carrier scattering and Auger recombination in the mid-region serve to reduce the value of L_a at injection levels above $10^{17}/cm^3$. Thus even if diode operation can be characterized as "short" at low current densities, it invariably shifts to "long" at high current densities.
3. Recombination in the end regions reduces the fraction of the forward current that is available for conductivity modulation of the mid-region. This, in turn, also results in an increase in the voltage drop V_M. From (3.105), it is seen that this effect can be reduced by lowering the geometric mean of J_{ns} and J_{ps}. Thus an improvement can result if either one (or both) of these saturation current densities can be lowered. This, in turn, requires that the p^+- and/or n^+-regions be heavily doped, but still preserve a long lifetime. To some extent these requirements are conflicting. In fact, practical experience with present-day technologies indicates that each of these terms is relatively independent of the process (diffusion or alloying) by which it is made. Typical observed values of J_{ns} and J_{ps} range from 1×10^{-13} to 4×10^{-13} A/cm^2, with J_{ns} usually smaller than J_{ps} by a factor of 1.5 to 2.
4. Finally, both carrier-carrier scatter and Auger processes set an upper limit to the diffusion length in the i-region. Thus at high current densities, the improvement that can be made by reducing recombination in the end regions is not as large as that predicted by (3.105). This is indicated in Fig. 3.10, where the calculated voltage drop at 1000 A/cm^2 is plotted as a function of i-region width for a number of different average values of saturation current density $J_s (\cong J_{ps} \cong J_{ns})$. The data points in this figure correspond to actual devices, made by a number of different technologies.

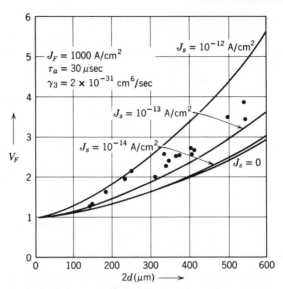

Fig. 3.10 The effect of J_s on the voltage drop. Copyright © 1975 by *Solid State Electronics*. Reprinted with permission from Burtscher et. al. [15].

5. Factors 1 through 4 apply for all p^+-i-n^+ diodes. In low voltage structures, however, it is possible to design devices whose d/L_a ratio remains small at high current densities. Thus these devices can behave as short diodes, even at current densities in excess of $1000 \text{ A}/\text{cm}^2$. High voltage structures, on the other hand, are invariably "long" at these high current densities.

Figure 3.11 depicts the results of calculations [15] on high voltage devices with $d = 300$, $\tau_a = 30$ µsec, showing the effects of including these different terms. Figure 3.11a considers only recombination in the mid- and end-regions, Fig. 3.11b includes the effect of carrier-carrier scattering, and Fig. 3.11c includes Auger recombination as well. The steady worsening of device characteristics is revealed in this progression, as well as the steadily reduced role of end-region recombination as other effects are included. These results serve to reinforce the conclusion that Auger recombination and carrier-carrier scattering set a limit on device performance, which means that reductions in J_{ps} and J_{ns} are not as useful as predicted by the theory.

Computer-aided studies of the forward-biased diffused junction p^+-i-n^+ diode have also been made [16]. It has been shown that the electrical base width in these structures widens under high level injection conditions, by

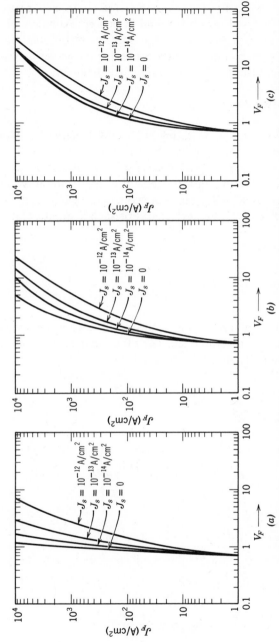

Fig. 3.11 Voltage versus current characteristics of p^+-i-n^+ diodes. Copyright © 1975 by Solid State Electronics. Reprinted with permission from Burtscher et. al. [15].

127

penetrating into the diffused region. Now, the minority carrier current injected into the diffused region is found to be approximately proportional to n_{+d}, and not to n_{+d}^2. As a result, the onset of recombination effects in the end regions is not as rapid, with a consequently slower deterioration of the device characteristics in the medium current range.

3.3.2   Reverse Recovery

Consider a p^+-ν-n^+ diode, operating at current densities for which recombination in the end regions can be ignored. Upon removal of forward bias, the charge stored in the ν-region decays by recombination to its thermal equilibrium value. During this process, the open circuit voltage is approximately given by $V_L + V_R$, since the voltage drop across the mid-region is relatively small and can be neglected. For such a situation, it was shown in (3.88) that

$$p_{-d} n_{+d} = n_i^2 \exp\left[\frac{q(V_L + V_R)}{kT} \right] \tag{3.111}$$

Let us assume that the diode is in high level operation before the current through the structure is terminated. Let n be the excess carrier concentration in the ν-region. Then

$$n = p_{-d} = n_{+d} \gg N_\nu^+ \tag{3.112}$$

so that

$$V_L + V_R = \left(\frac{kT}{q} \right) \ln\left(\frac{n^2}{n_i^2} \right) \tag{3.113}$$

and

$$\frac{d(V_L + V_R)}{dt} = \frac{2kT}{q} \frac{dn/dt}{n} \tag{3.114}$$

In high level operation, the lifetime is defined by

$$\tau_{\text{high}} = -\frac{dn/dt}{n} \tag{3.115}$$

Combining with (3.114), the open circuit voltage across the diode terminals is described by

$$\frac{d(V_L + V_R)}{dt} = -\frac{kT}{q}\frac{2}{\tau_{\text{high}}} \tag{3.116}$$

Eventually the carrier density in the device decays to the point at which the injected majority carrier concentration becomes less than the background concentration N_ν^+. The device is now in low level operation, and

$$p_{-d}n_{+d} \cong p_{-d}N_\nu^+ \tag{3.117}$$

Making this substitution in (3.111) and writing the lifetime as τ_{low}, gives

$$\frac{d(V_L + V_R)}{dt} = -\frac{kT}{q}\frac{1}{\tau_{\text{low}}} \tag{3.118}$$

In practice, these relationships hold over many decades of injected carrier density, providing a basis for the measurement of the lifetime in the ν-region by an extremely simple and rapid technique [17].

From a circuit point of view, the excess charge in the ν-region must fully decay before the device becomes fully recovered, in about 7 to 10τ. This presents a problem to the device designer, since a long lifetime is necessary to obtain a low ON state voltage. In a practical circuit situation, the forward-biased diode can be turned off in a much shorter time by the application of a reverse voltage V_{REV} in series with a large resistor R. Immediately, the forward voltage across the diode falls by a small value because of the reversal of the ohmic component of the drop across the mid-region. This voltage drop is $(1 + J_R/J_F)V_M$, and it can be ignored for all practical purposes. At the same time, the carrier concentration alters in accordance with Fig. 3.12; initially the full reverse current V_{REV}/R flows through the diode. This current consists of holes swept out of the mid-region from right to left, and of electrons swept out from left to right. Full reverse current persists until a space charge region begins to form at one of the two junctions. Since current flow is by diffusion, it follows that the current density is given by

$$qD_p\frac{dp}{dx}\bigg|_{x=-d} = -qD_n\frac{dn}{dx}\bigg|_{x=+d} \tag{3.119}$$

For silicon, $\mu_n > \mu_p$, so that

$$\frac{dp}{dx}\bigg|_{x=-d} > -\frac{dn}{dx}\bigg|_{x=+d} \tag{3.120}$$

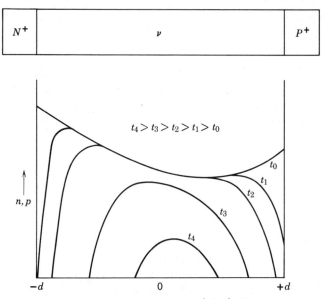

Fig. 3.12   Carrier decay in a p^+-i-n^+ diode.

Thus if t_{OL} and t_{OR} are the times required to establish space charge regions at the left- and right-hand junctions, respectively, then $t_{OL} < t_{OR}$, and the space charge zone starts to develop initially at the p^+-ν junction.

Up to this point, the diode is still forward biased. Now, however, reverse voltage across the diode builds up by the formation of space charge at the ends of the mid-region, until it reaches a value V_{REV}. Simultaneously, the device becomes unable to support the full reverse current V_{REV}/R, which eventually falls off to zero (since leakage current can be ignored for silicon devices). The current and voltage wave forms during this recovery phase appear in Fig. 3.13.

Detailed computations of t_{OL} and t_{OR} have been made [8] by solving the time-varying form of the continuity and diffusion equations. However physical insight into the process can be obtained by the use of charge control methods [18]. For simplicity, an *intrinsic* mid-region is assumed in this analysis.

In Fig. 3.14 consider the removal of charge at time $t = t_{OL}$, where the charges of Q_L and Q_R have been swept out on the left- and right-hand sides, respectively. Then

$$Q_L \cong bQ_R \qquad (3.121)$$

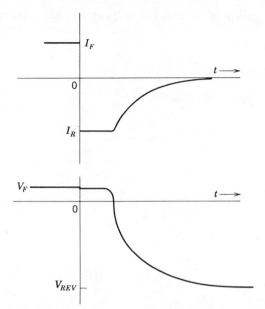

Fig. 3.13 Reverse recovery characteristics of a p^+-i-n^+ diode.

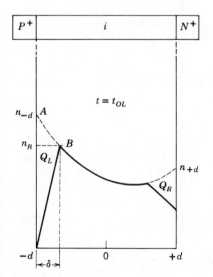

Fig. 3.14 Device behavior at $t = t_{OL}$.

If the curved section of the catenary AB can be approximated by a straight line, then

$$Q_L \cong \frac{q\delta n_{-d}}{2}$$
(3.122)

Furthermore, the reverse current

$$J_R = \frac{2qD_p n_B}{\delta} \cong \frac{2qD_p n_{-d}}{\delta}$$
(3.123)

since δ is typically much smaller* than d. Finally from charge control,

$$J_R t_{OL} = Q_L + Q_R$$
(3.124)

Combining (3.121) to (3.124) gives

$$t_{OL} = q^2 D_p \left(1 + \frac{1}{b}\right)\left(\frac{n_{-d}}{J_R}\right)^2$$
(3.125)

Here n_{-d} can be obtained by substitution in (3.75). Solving, we have

$$\frac{t_{OL}}{\tau_a} = \left(\frac{J_F}{J_R}\right)^2 \frac{(1 + 1/b)^2}{8} \coth\left(\frac{d}{L_a}\right)\left[1 + B\tanh^2\left(\frac{d}{L_a}\right)\right]^2$$
(3.126a)

$$\frac{t_{OR}}{\tau_a} = \left(\frac{J_F}{J_R}\right)^2 \frac{(1 + b)^2}{8} \coth\left(\frac{d}{L_a}\right)\left[1 - B\tanh^2\left(\frac{d}{L_a}\right)\right]^2$$
(3.126b)

For example,

$$\frac{t_{OL}}{\tau_a} = 0.64\left(\frac{J_F}{J_R}\right)^2$$
(3.127a)

$$\frac{t_{OR}}{\tau_a} = 1.73\left(\frac{J_F}{J_R}\right)^2$$
(3.127b)

for $d/L_a = 1$, in close agreement to that given by the detailed theory [8]. Note that $t_{OL} < t_{OR}$, as indicated earlier.

Reverse voltage across the diode builds up by the formation of space charge at the two ends of the i-region (Fig. 3.15). The edges of the two

*See Problem 3.4.

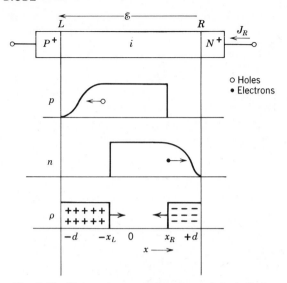

Fig. 3.15   Space charge regions in a p^+-i-n^+ diode.

space charge regions proceed to sweep inward toward each other in time, as the reverse voltage increases. Thus the space charge is built up from movable charge carriers alone.* Let $-x_L$ and x_R define these edges at time t. On the p^+-side, the hole space charge density ρ_L is given by the hole concentration, so that

$$\frac{d\mathscr{E}_L}{dx} = \frac{qp}{\varepsilon\varepsilon_0} \qquad (3.128)$$

where

$$\mathscr{E}_L = 0 \qquad \text{at} \quad x = -x_L \qquad (3.129)$$

The diffusion current can be neglected in the swept-out zone. Thus the reverse current density is given (assuming a constant reverse current) by

$$J_R = -q\mu_p p \mathscr{E}_L$$

Solution of these equations gives

$$\mathscr{E}_L = -\left[\frac{2}{\varepsilon\varepsilon_0\mu_p} J_R (-x - x_L) \right]^{1/2} \qquad (3.130)$$

---

*This is only true for an intrinsic mid-region, assumed here. The role of the fixed charges is considered in Section 3.3.2.1.

and

$$\rho_L = \frac{1}{q}\left(\frac{\varepsilon\varepsilon_0}{2\mu_p} \frac{J_R}{-x-x_L} \right)^{1/2} \tag{3.131}$$

The voltage supported by the space charge layer V_L is obtained by integrating the electric field from the junction to the depleting layer edge and is given by

$$V_{L,sc} = \frac{2}{3}\left(\frac{2J_R}{\varepsilon\varepsilon_0\mu_p} \right)^{1/2} (d-x_L)^{3/2} \tag{3.132}$$

Similar equations hold at the right-hand junction.

The time variation of $V_{L,sc}$ and $V_{R,sc}$ can be obtained by application of charge control principles. Thus for the left-hand junction, assuming equal electron and hole mobilities, we have

$$J_R\,dt = qn^*\,d\,(-x_L) \tag{3.133a}$$

and for the right-hand junction,

$$J_R\,dt = qn^*\,d\,(x_R) \tag{3.133b}$$

so that

$$\frac{dx_L}{dt} = \frac{dx_R}{dt} = \frac{J_R}{qn^*} \tag{3.134}$$

Calculations [8] that include the effect of dissimilar mobilities give

$$\frac{dx_L}{dt} = \frac{b}{b+1}\frac{J_R}{qn^*} \tag{3.135a}$$

$$\frac{dx_R}{dt} = \frac{1}{b+1}\frac{J_R}{qn^*} \tag{3.135b}$$

Thus the space charge region grows faster on the p^+-i side by a factor of 3 for silicon devices.

Using these relations, and noting that the boundaries begin moving inwards at t_{OL} and t_{OR}, respectively, gives the voltage across the space charge region as

$$V_{L,sc} = \frac{2}{3}\left(\frac{2}{\varepsilon\varepsilon_0 q^3 \mu_p}\right)^{1/2} \frac{b^2}{(b+1)^{3/2}} \frac{J_R^2}{(n*)^{3/2}}(t-t_{OL})^{3/2} \quad (3.136a)$$

$$V_{R,sc} = \frac{2}{3}\left(\frac{2}{\varepsilon\varepsilon_0 q^3 \mu_n}\right)^{1/2} \frac{1}{(b+1)^{3/2}} \frac{J_R^2}{(n*)^{3/2}}(t-t_{OR})^{3/2} \quad (3.136b)$$

Since $b = 3$ for silicon, the rate of voltage buildup on the p^+-side ($V_{L,sc}$) is 5.2 times as large as the rate of buildup on the n^+-side. Furthermore, since t_{OL} occurs before t_{OR}, the actual buildup of reverse voltage is almost completely across the p^+-i junction. Finally, this buildup occurs as a function of the three halves power of time, for $t \geqslant t_{OL}$.

The reverse recovery process, as defined by (3.136), persists until both swept regions meet. At this point the device enters a new phase during which the positive and negative space charges are annihilated. Annihilation proceeds *outward* from the point of their contacts, therefore, in effect, the two space charge regions recede, at a velocity approximately equal to the limiting velocity. This phase is of considerable importance in the device physics of TRAPATT oscillators [19]. However its duration is extremely short, and it can be neglected in a study of power semiconductor devices.

3.3.2.1 Mid-Region Doping Effects

The theory of the p^+-i-n^+ diode recovery can now be extended to include the case of finite doping (ν- or π-type) in the mid-region. Physically, the situation is unchanged at high reverse currents, where the carrier concentration exceeds the doping level of this region. Now, a space charge voltage builds up at the p^+- and n^+-junctions after times t_{OL} and t_{OR}, respectively, as before. Eventually, however, this condition is violated, and the two junctions behave differently. The junction comprising dissimilar conductivity types develops a space charge zone that expands as described, and the voltage drop across it is proportional to the square of its width. Since, however, the "junction" comprising similar conductivity types approaches ohmic behavior, the voltage drop across it is very small, and linearly proportional to the width of the swept-out zone. As a result, the voltage developed across the true junction (p^+-ν or n^+-π) is the dominant one, and the ohmic drop can soon be ignored. Thus the true junction (dissimilar conductivity type) dominates the reverse recovery characteristics.

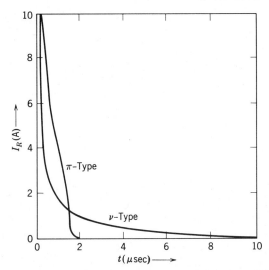

Fig. 3.16   Recovery characteristics of π- and ν-type diodes. Copyright © 1967 by the Institute of Electrical and Electronic Engineers, Inc. Reprinted with permission from Benda and Spenke [8].

The initial recovery phase, during which the full reverse current flows, terminates *earlier* on the p^+-junction side for both p^+-ν-n^+ and p^+-π-n^+ diodes. In addition, the space charge region develops on the p^+ side for the p^+-ν-n^+ diode, but on the n^+-side for the p^+-π-n^+ diode. Moreover, (3.134) indicates that the space charge region develops *faster* on the p^+-side than on the n^+-side, by a factor of 3. Consequently, the initial recovery process in a p^+-ν-n^+ diode is terminated considerably faster than for a p^+-π-n^+ diode with similar device parameters. On the other hand, both devices must sweep out the same amount of stored charge for full recovery, assuming similar dimensions and lifetimes for the mid-regions. As a result, the reverse current in the π-type device terminates abruptly, whereas that in a ν-type device tends to decay over a longer time interval. Thus the overall recovery transient of a π-type device is of shorter duration, as represented (Fig. 3.16) in the recovery characteristics [8] for two devices, made by identical processes, except for the mid-region resistivity type. Devices shown had a mid-region doping concentration of $7 \times 10^{14}/\text{cm}^3$, $d = 100$ μm, $\tau = 11$ μsec, $I_F = 1$ A and $I_R = 10$ A in each case.

3.3.2.2   Diffused Junction Effects

Either one or both end regions of a p^+-i-n^+ diode may be made by diffusion. Now the effect of an electric field in the heavily doped region must also be considered. Referring to Fig. 3.15, assume that the p^+-region is diffused. This sets up an electric field in the negative direction and helps in the transport of holes across this junction. Thus the recovery characteristic is aided by this electric field. A diffused n^+-region also results in an electric field in the negative direction which facilitates the removal of electrons at the right-hand junction. Again, the recovery characteristic is aided by this process. Finally, we note that this improvement in the recovery characteristic is independent of the dopant type of the mid-region. Thus a diode with one or more diffused junctions will have a somewhat faster recovery characteristic than its abrupt counterpart.

3.4 INSTABILITIES

Two types of instability are possible in a forward-biased diode. The first is occasionally seen in high voltage p^+-i-n^+ diodes, where the mid-region is long relative to the diffusion length. Here it is possible to obtain double injection and a forward characteristic of the type shown in Fig. 1.3. This negative resistance characteristic results in filamentary conduction at the point of turn-on and can lead to mesoplasma formation if the intrinsic temperature is exceeded in the filamentary region.

Lateral instability is purely thermal in origin. This second type of instability leads to the formation of a localized region with current density well in excess of the average value, until the intrinsic temperature is reached. This is followed by a very rapid increase of current and the growth of a destructive mesoplasma.

Thermal instability in forward-biased junction diodes can be investigated [20] by way of the equation governing their characteristics. Thus for n^+-p diodes, we have

$$I \cong \frac{qAD_n n_i^2}{\bar{p}_P L_n \tanh(W/L_n)} e^{qV_A/kT} \tag{3.137}$$

Where the major temperature dependencies are carried in the exponential terms associated with n_i^2 and with the applied voltage. But

$$n_i^2 = N_v N_c e^{-E_g/kT} \tag{3.138}$$

where N_v and N_c are the density of states at the valence and conduction band edges, respectively. Moreover, the energy gap E_g is given by [21]

$$E_g = (1.21 - 3.6 \times 10^{-4} T)\,\text{eV} \tag{3.139}$$

where T is the temperature in degrees Kelvin.

Combining these equations and replacing all terms outside the exponential by I', the forward current can be written as

$$I \cong I' \exp\left(\frac{qV_A - E_g}{kT}\right) \tag{3.140}$$

For any constant forward current, therefore,

$$\left(\frac{dV_A}{dT}\right)_{I = \text{const}} \cong \frac{1}{q}\frac{dE_g}{dT} + \frac{k}{q}\ln\frac{I}{I'} \tag{3.141}$$

Typically, this quantity is relatively constant at about -2 mV/$^\circ$C for silicon diodes in the temperature range between 30 and 200°C. Thus for any constant forward current, the voltage across a diode falls about 2 mV per $^\circ$C of temperature rise. Equation 3.140 is often written in a more useful form, in terms of the actual change in temperature, as

$$I = I^* \exp\left[\frac{q(V_A + \phi\Delta T)}{kT}\right] \tag{3.142}$$

where $\phi = 2$ mV/$^\circ$C and I^*, a temperature independent term, is constant for any given diode.

At sufficiently high power levels, the improved conduction properties of a diode can cause the current to build up locally in one region and to decrease in another. This can be seen by considering two regions of a diode as separate but identical devices A and B, placed in parallel and fed from a constant current source. Any increase of temperature in, say, device A, results in its improved conduction characteristics and lowers the forward voltage across both the devices. Thus the forward current in device B falls, causing further shift of current over to device A. The process is a regenerative one—that is, one device continues to rob current from the other—resulting in nonuniform current flow.

Lateral thermal instability effects are considerably more serious in transistors, where the power dissipated at the collector-base junction is much greater than that dissipated in a forward-biased diode. A more detailed discussion of this problem is deferred to Chapter 4.

3.5 REFERENCES

1. J. Moll, *Physics of Semiconductors*, McGraw Hill Book Co., New York, 1964.
2. S. K. Ghandhi, *The Theory and Practice of Microelectronics*, John Wiley & Sons, New York, 1968.

3. J. L. Moll and I. M. Ross, "Dependence of Transistor Parameters on the Distribution of Base Layer Resistivity," *Proc. IRE*, **44**, No. 1, pp. 72–78 (1956).

4. A. A. Vol'fson and V. K. Subashiev, "Fundamental Absorption Edge of Silicon Heavily Doped with Donor or Acceptor Impurities," *Sov. Phys.—Semicond.*, **1**, pp. 327–332 (September 1967).

5. R. P. Mertens et al., "Calculation of the Emitter Efficiency of Bipolar Transistors," *IEEE Trans. Electron Devices*, **ED-20**, No. 9, pp. 772–778 (1973).

6. H. J. J. DeMan, "The Influence of Heavy Doping on the Emitter Efficiency of a Bipolar Transistor," *IEEE Trans. on Electron Devices*, **ED-18**, No. 10, pp. 833–835 (1971).

7. R. N. Hall, "Power Rectifiers and Transistors," *Proc. IRE*, **40**, No. 11, pp. 1512–1518 (1952).

8. H. Benda and E. Spenke, "Reverse Recovery Processes in Silicon Power Rectifiers," *Proc. IEEE*, **55**, No. 8, pp. 1331–1354 (1967).

9. S. C. Choo, "Effect of Carrier Lifetime on the Forward Characteristics of High Power Devices," *IEEE Trans. Electron Devices*, **ED-17**, No. 9, pp. 647–652 (1970).

10. F. Dannhauser and J. Krause, "Die Raümliche Verteilung der Rekombination in Legierten Silizium *p-s-n* Gleichrichtern Bei Belastung in Durchlassrichtung," *Solid State Electron.*, **16**, No. 8, pp. 861–873 (1973).

11. N. H. Fletcher, "The High Current Limit for Semiconductor Junction Devices," *Proc. IRE*, **45**, No. 6, pp. 862–872 (1957).

12. J. Krause, "Auger-Rekombination im Mittelgebiet Durchlassbelaster Silizium-Gleichrichter und-Thyristoren," *Solid State Electron.*, **17**, No. 5, pp. 427–429 (1974).

13. N. R. Howard and G. W. Johnson, "PIN Silicon Diodes at High Forward Current Densities," *Solid State Electron.*, **8**, No. 3, pp. 275–284 (1965).

14. A. Herlet, "The Forward Characteristic of Silicon Power Rectifiers at High Current Densities," *Solid State Electron.*, **11**, No. 8, pp. 717–742 (1968).

15. J. Burtscher et al., "Die Rekombination in Thyristoren und Gleichrichtern aus Silizium," *Solid State Electron.*, **18**, No. 1, pp. 35–63 (1975).

16. S. C. Choo, "Theory of a Forward-Biased Diffused-Junction *P-L-N* Rectifier—Part 1. Exact Numerical Solutions," *IEEE Trans. Electron Devices*, **ED-19**, No. 8, pp. 954–966 (1972).

17. L. W. Davies, "The Use of *P-L-N* Structures in Investigations of Transient Recombination from High Injection Levels in Semiconductors," *Proc. IEEE*, **51**, No. 11, pp. 1637–1642 (1963).

18. P. E. Cottrell, private communication.

19. R. C. Varshney and D. J. Roulston, "Turn-Off Transient Behavior of *p-i-n* Diodes," *Solid State Electron.*, **14**, No. 8, pp. 735–745 (1971).

20. R. H. Winkler, "Thermal Properties of High Power Transistors," *IEEE Trans. Electron Devices*, **ED-14**, No. 5, pp. 260–263 (May 1967).

21. C. Popescu, "Self-Heating and Thermal Runaway Phenomena in Semiconductor Devices," *Solid State Electron.*, **13**, No. 4, pp. 441–450 (1970).

3.6 PROBLEMS

1. A p^+-n diode has a low level lifetime of 10 μsec and an n-region doping concentration of $5 \times 10^{14}/\text{cm}^3$. Compute the components due

to diffusion and recombination for currents from 0 to 10 mA. Assume a cross section of $1\,\text{mm} \times 1\,\text{mm}$, and a depletion layer width of $1\,\mu\text{m}$ under forward bias conditions.

2. The diode of Problem 1 is biased to pass a forward current of 10 A. At time t_0, the voltage across the diode is reversed. Calculate the maximum peak current that will flow at the instant of reversal. Assume no resistance in the external circuit.

3. Consider the situation depicted in Fig. 3.2, except that a capacitance C is shunted across R. Compute the recovery characteristic that would be obtained for this situation, if $CR = \tau_F$.

4. Compute the carrier concentration in the mid-region for the device described in Fig. 3.11a, for the case of $J_s = 0$, and for current densities of 1, 100, and 3000 A/cm².

5. Repeat Problem 4 for $J_s = 10^{-12}$ A/cm².

6. A p^+-i-n^+ diode has a d/L_a ratio of 2. Compute the ratio of t_{OL}/t_{OR} as defined in Section 3.2.

7. A p^+-i-n^+ diode is made with a mid-region doping of 10^{14}/cm³ and $d = L_a = 150$ μm. Calculate the value of δ, as defined in (3.122), for a reverse current density of 100 A/cm². Compare this value with d.

8. Investigate the nature of the approximations made in developing (3.142) from (3.140).

9. Using charge control principles, develop (3.135a) and (3.135b).

The Transistor

CONTENTS

DOMINATING THE BEHAVIOR of power transistors is the need to support a high reverse voltage across the collector-base junction. This requirement affects the design of the device, and even its structural configuration in many cases. High starting resistivity, deep diffusions, and wide bases are typical consequences of this requirement. These factors play a significant role in determining a transistor's dc and high frequency performance.

This chapter does not attempt to cover the basic theory of transistor operation, which has been treated in many texts. Rather, it supplies details relating to areas in which the power device differs from its small signal counterpart.

4.1 DEVICE STRUCTURES

The double diffused epitaxial process [1] is almost universally chosen for fabricating small signal, low voltage transistors. For power transistors, however, other processes can more readily meet the high breakdown voltage requirement. Thus the doping profile of a commonly available high voltage transistor (Fig. 4.1a) consists of deep diffusions simultaneously made into both sides of a slice of lightly doped material. This transistor, made by a single diffusion, has many design-related problems that are not present in conventional small signal devices. For example,

1. The collector junction is diffused into a uniform and lightly doped slice of starting material. As a consequence, this type of transistor is highly susceptible to failure by punchthrough before avalanche breakdown conditions are reached at the collector-base junction.
2. A wide base region must be used, to attain a high punchthrough voltage. Furthermore, the base width is a function of slice thickness, thus varies with the starting material. Consequently the role of the base transport factor is important in determining both the current gain and the gain-bandwidth product of this device.

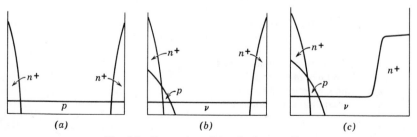

Fig. 4.1   Power transistor doping profiles.

3. The emitter junction is deep diffused, since the device is made by a single diffusion step; this will be reflected in a slightly reduced injection efficiency, hence in relatively low gain.

A significant improvement in the breakdown voltage can be obtained with a n^+-p-v-n^+ structure whose doping profile is shown in Fig. 4.1b. Here the additional reverse voltage that can be supported because of the v-region is approximately equal to the breakdown voltage of a reverse-biased p^+-v-n^+ diode of comparable width, as given by (2.37). Furthermore, since the movement of the collector depletion layer into the diffused base is now inhibited by the grading, failure by punchthrough seldom occurs in this device.

In normal operation, the v-region is fully depleted, and the device behaves like a double diffused n^+-p-n^+ transistor, with a relatively narrow (5–10 μm) base width. The frequency response of the structure is thus considerably superior to that of the single diffused transistor. As a result of these many advantages, the n^+-p-v-n^+ structure is the basic modern high voltage power transistor.

In an n^+-p-v-n^+ transistor made by the double diffused epitaxial process (Fig. 4.1c), the v-region is grown epitaxially on an n^+-substrate and is followed by base and emitter diffusions. This process, identical to that used for modern small signal devices and microcircuits, has significant advantages over the last from the point of view of device fabrication (see Chapter 6). Unfortunately, it is extremely difficult to grow epitaxial layers having resistivities as high as can be obtained in bulk silicon; thus its reverse breakdown voltage is lower than that of n^+-p-v-n^+ devices.

A double epitaxial structure can be used [2] to obtain an improvement in punchthrough voltage. This structure, n^+-p-π-v-n^+, is presented in Fig. 4.2, together with its electric field configuration at breakdown. Also shown is an n^+-p-v-n^+ structure with the same width of lightly doped region. For comparison, the doping level in all the lightly doped regions is assumed equal, and the peak electric field at breakdown is assumed to be the same for both devices. Then, from geometrical considerations,

$$BV_{np\pi vn} - BV_{npvn} \cong \frac{qNW^2}{4\varepsilon\varepsilon_0} \tag{4.1}$$

where N is the doping level in both the lightly doped regions and W represents their combined widths. This equation indicates that the improvement in BV is proportional to N. This technique is thus particularly useful for epitaxial devices in which it is difficult to achieve low doping for this region.

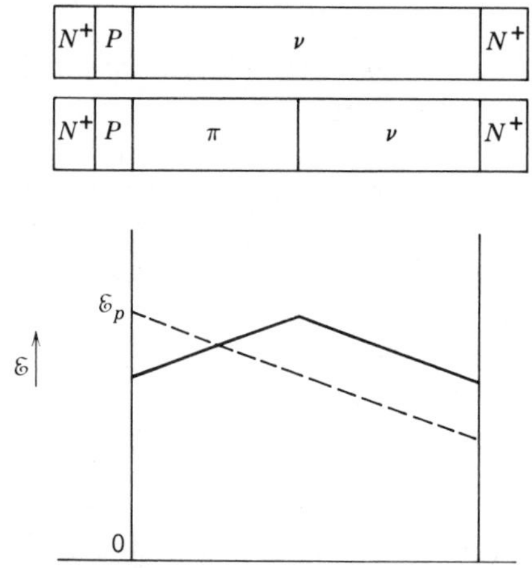

Fig. 4.2 The double epitaxial transistor.

4.2 OUTPUT CHARACTERISTICS

The output characteristics for a modern diffused n^+-p-n^+ transistor are given in Fig. 4.3a. For a specific base current, the device characteristic exhibits* an ohmic saturation region 0–A, followed by an active region A–B. The collector-base junction is forward biased over the region 0–A and has a slope equal to the ohmic resistance of the n^+-collector body ($\ll 1$ ohm). The forward bias decreases from 0 to A, at which point it becomes zero. From A to B, the collector-base junction is reverse biased; the device is now in its active region, with a dynamic resistance of many kilohms.

Figure 4.3b shows the output characteristics for a typical high voltage n^+-p-ν-n^+ transistor having a wide ν-region of resistance R_ν. Here, for a base current of I_{B1}, the characteristic exhibits saturation from 0 to A′, with a resistance equal to that of the n^+-collector. From A′ to A, however, the slope changes until the device enters its active region at A. The region A′–A is known as the quasi-saturation region. From A to B, the collector-base junction is reverse biased, and the device is again in its active region.

The quasi-saturation characteristics of an n^+-p-ν-n^+ transistor can be understood [3–5] by noting that the collector-base junction becomes less forward biased from 0–A′–A, is zero biased at A, and eventually becomes reverse biased beyond A. In the region 0–A′, the p-ν junction is heavily forward biased, which ensures that the ν-region is filled with injected holes. In addition, charge neutrality requires that the electron concentration in this region be approximately equal to the hole concentration. Thus the

*A small offset voltage, typically under 100 mV, has been ignored for both devices in this figure.

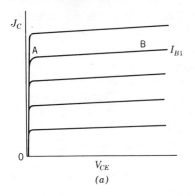

(a)

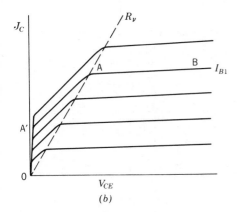

(b)   **Fig. 4.3   Saturation characteristics.**

ν-region becomes heavily conductivity modulated and supports a very small potential drop.

The p-ν junction is less heavily biased between A′ and A. Now the ν-region becomes conductivity modulated over only part of its length, which serves to increase its resistance. Eventually at A, the voltage across the p-ν junctions becomes zero, and hole injection ceases. The effective collector resistance of the device is now given by the resistance of the ν-region, and the body resistance of the n^+-collector can be ignored. The dashed line 0A thus has a slope of R_ν (Fig. 4.3b).

From A onward the p-ν junction is reverse biased, and the collector-base depletion layer fills the ν-region. The device is now in its active mode. The line 0A divides the device characteristics into two regions of operation. To its right the device is active, and operation is similar to that in a normal transistor. To its left, however, the device operates in a quasi-saturated mode.

Some of the features of high voltage transistor operation in quasi-saturation can be summarized, as follows:

1. The ON state voltage of this device is seen to be impaired, together with its power handling capability, under steady state conduction.
2. The current gain is seen to fall off at low voltages. Thus it needs more base drive to operate as an effective switch than an n^+-p-n^+ transistor.
3. For any given supply voltage, the current gain falls off rapidly at high levels, as the device moves into its quasi-saturation region.
4. Although not apparent from these static characteristics, it will be shown that the turn-on time is increased as device operation extends into the quasi-saturation region. Thus the power dissipation during the turn-on transient may act to limit its power handling capability.

Figure 4.4a shows the n^+-p-ν-n^+ transistor with its p-ν junction forward biased. For this situation, the ν-region is filled with injected holes, with a

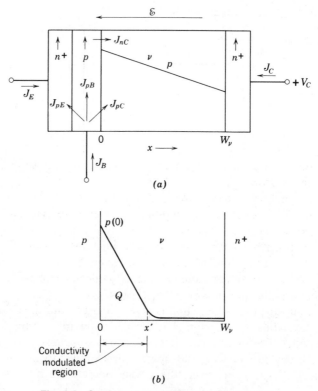

Fig. 4.4   Carrier concentrations in the ν-region.

concentration that is well in excess of the background. The electron concentration in the ν-region is given by

$$n \cong p + N_\nu \qquad (4.2)$$

so that both carriers are in essentially equal number. The region is thus heavily conductivity modulated.

Electrons in the ν-region are due in part to the transport of minority carriers from the emitter to the collector, by transistor action. Device design requires that their concentration be well below N_ν. Thus nearly all the electron concentration in the ν-region is due to the charge neutrality requirement and is supplied by injection across the n^+-ν barrier.

The hole and electron current densities in the collector are given by

$$J_{pC} = q\mu_p p \mathcal{E} - qD_p \frac{dp}{dx} \qquad (4.3)$$

and

$$J_{nC} = q\mu_n (p + N_\nu) \mathcal{E} + qD_n \left(\frac{dn}{dx} \right) \qquad (4.4)$$

respectively. The total collector current density $J_C = -(J_{pC} + J_{nC})$ is a negative quantity for the sign convention shown in Fig. 4.4.

The electric field in the collector region, obtained by solving (4.3), is given by

$$\mathcal{E} = \frac{kT}{q} \frac{1}{p} \frac{dp}{dx} + \frac{J_{pC}}{q\mu_p p} \qquad (4.5a)$$

Note that it has two components. The first term is due to the change in hole concentration, caused by forward injection of the p^+-ν diode. The second term is due to the ohmic drop of J_{pC} across the ν-region. Now J_{pC} is the hole current density provided by the base, to sustain recombination in the ν-region. Thus at most it is equal to the base current density minus that part necessary for recombination in the base and for hole injection into the emitter (see Fig. 4.4a). Finally, our interest is in the regime where the transistor has a significant current gain. To an excellent approximation, therefore, $J_{pC} \cong 0$, and the electric field is given by

$$\mathcal{E} \cong \frac{kT}{q} \frac{1}{p} \frac{dp}{dx} \qquad (4.5b)$$

At first glance it would seem unusual that $J_{pC} \cong 0$, even though it provides the holes needed for conductivity modulation of the ν-region.

This apparent anomaly can be explained, however, by noting that the situation is very similar to what is encountered in the i-region of a p^+-i-n^+ diode. In both cases, the function of this hole component is to provide for recombination in the mid-region, so that its magnitude is inversely proportional to the mid-region lifetime. See, for example, (3.65) for the injected current density in a p^+-i-n^+ diode.

Combining (4.4) with (4.5b) and noting that $dn/dx = dp/dx$, we obtain

$$J_C \cong -J_{nC} = 2qD_n \left(1 + \frac{N_\nu}{2p} \right) \frac{dp}{dx} \tag{4.6}$$

Solution of this equation gives

$$p(x) = p(0) - \frac{J_C x}{2qD_n} + \frac{N_\nu}{2} \ln \frac{p(0)}{p(x)} \tag{4.7}$$

where $p(0)$ is the hole concentration at the edge of the depletion layer. The last term in this equation can be ignored for device operation in saturation, since $p \gg N_\nu$. Thus the hole concentration is a linearly decreasing function of distance, as in Fig. 4.4a.

Figure 4.4b shows the hole concentration for a bias point in the A'–A region, as depicted in Fig. 4.3b. Again, the curve is linearly decreasing up to the point x', where

$$x' \cong \frac{2qD_n p(0)}{J_C} \tag{4.8}$$

For $x' \leqslant x \leqslant W_\nu$, the hole concentration is given by its equilibrium value and is equal to n_i^2/N_ν. At $x = x'$, (4.7) reduces to

$$p(x') = \frac{N_\nu}{2} \ln \frac{p(0)}{p(x')} \tag{4.9}$$

A closed form solution of this equation is not possible; however $p(x')$ can be approximated by N_ν for a wide range of values of both N_ν and $p(0)$. Making this approximation, the potential drop in the conductivity modulated part of the ν-region is given by $V_{0x'}$, where

$$V_{0x'} = -\int_0^{x'} \mathscr{E} \, dx = -\frac{kT}{q} \int_{p(0)}^{p(x')} \frac{dp(x)}{p(x)} \cong \frac{kT}{q} \ln \frac{p(0)}{N_\nu} \tag{4.10}$$

This value, typically 100–200 mV, can be ignored during device operation in quasi-saturation.

The potential drop across the unmodulated part of the ν-region is caused by collector current flow through its ohmic resistance and is given by

$$V = \frac{J_C(W_\nu - x')}{q\mu_n N_\nu} \qquad (4.11)$$

since $J_C \cong -J_{nC}$. This is approximately equal to the collector-emitter voltage of the n^+-p-ν-n^+ transistor in its ON condition, for operation in quasi-saturation.

The injected hole concentration $p(0)$ can be determined from charge control considerations. Referring to Fig. 4.4b, and writing τ_ν as the hole lifetime in the ν-region, the stored charge is given by Q, such that

$$\frac{Q}{A} = J_{pC}\tau_\nu \cong \frac{qp(0)x'}{2} \qquad (4.12)$$

where A is the cross section area of the ν-region. But

$$\frac{dp}{dx} \cong -\frac{p(0)}{x'} = -\frac{J_C}{2qD_n} \qquad (4.13)$$

so that

$$p(0) = \left(\frac{J_C J_{pC}\tau_\nu}{q^2 D_n}\right)^{1/2} \qquad (4.14)$$

and

$$x' = 2\left(\frac{J_{pC}\tau_\nu D_n}{J_C}\right)^{1/2} \qquad (4.15)$$

Modulation of the p-base region has been ignored here, since its background concentration is much higher than N_ν.

The onset of quasi-saturation occurs at the point A′ in Fig. 4.3b. Writing this current density as J_T and substituting $x = W_\nu$ in (4.7) gives this value as

$$J_T = \frac{2qD_n p(0)}{W_\nu} \qquad (4.16)$$

Thus the current density at which quasi-saturation begins is directly proportional to the injected hole density $p(0)$.

From (4.14) and (4.16) it appears that for any specific lifetime (corresponding to a particular manufacturing process), J_T can be increased by increasing J_{pC}, even to the point at which the static characteristics of the n^+-p-ν-n^+ transistor exhibit no sign of quasi-saturation. Unfortunately, this is not the case for the following reasons:

1. Increase of J_{pC} is accomplished by raising the doping level in the base. This results in a fall in the common emitter gain β_0, necessitating additional base current drive for circuit operation. To a crude approximation, J_T varies as $(\beta_0)^{-1/2}$; therefore the saturation performance for any given base drive worsens as the base doping level is increased.
2. Even if all visible signs of quasi-saturation are eliminated in the static characteristic,* 0–A still divides this characteristic into two regimes corresponding to active and quasi-saturation operation. Thus quasi-saturation effects are observed whenever the operating point of the device moves over these two regions, as occurs during the turn-on process.

Figure 4.5a shows an output characteristic for three types of transistors operating from a voltage V_{CC} through a load resistance R_L, where 0A′AB is the characteristic for a n^+-p-ν-n^+ transistor that exhibits quasi-saturation effects and 0A″B is the characteristic of an n^+-p-n^+ transistor having no ν-region. The line 0A″B also represents the characteristic for an n^+-p-ν-n^+ transistor designed so that quasi-saturation is not apparent in the static characteristic. The turn-on transients for these three devices are now compared [3], on the assumption of a step of input base current I_{B1}. Curve 1 in Fig. 4.5b shows V_{CE} for the n^+-p-n^+ transistor and illustrates a conventional turn-on transient, as well as the ON voltage V_{CE1}. This transient can be characterized by a single time constant [1].

Curve 2 shows V_{CE} for an n^+-p-ν-n^+ transistor whose output characteristic is indicated by 0A′AB. For this device, the turn-on is fast until the collector voltage reaches V_{CE3}, at which point the device goes into its quasi-saturation regime. Operation is now considerably slower, with a steady state ON voltage of V_{CE2}. Thus the turn-on process must be characterized by two time constants.

Curve 3 shows V_{CE} for an n^+-p-ν-n^+ transistor whose output characteristic is indicated by 0A″B. Here, after a fast transient to V_{CE3}, the device

---

*This can be accomplished by making J_T equal to or greater than the collector current density when the device is in its active region.

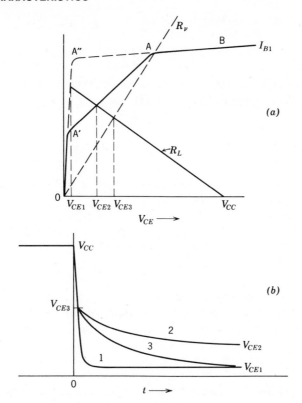

Fig. 4.5   Turn-on characteristics of the n^+-p-ν-n^+ transistor.

again goes into quasi-saturation where its operation is slower. Eventually a steady state ON condition is reached, at a voltage given by V_{CE1}. Again, two time constants are required to describe this process.

An additional point is worth noting. The amount of time that the device spends in its quasi-saturation regime in case 3 is greater than that for case 2. Consequently the turn-on process takes longer to accomplish for the former situation.

Charge buildup in the ν-region during the turn-on process can be determined by solving the charge control equation for $J_{pC}(t)$. Thus

$$J_{pC}(t)A = \frac{dQ(t)}{dt} + \frac{Q(t)}{\tau_\nu} \tag{4.17}$$

where A is the cross-section area of the ν-region and $Q(t)$ is the excess hole charge. In steady state, this charge is $J_{pC}\tau_\nu A$, so that

$$Q(t) = J_{pC}\tau_\nu A (1 - e^{-t/\tau_\nu}) \tag{4.18}$$

The modulated part of the ν-region expands with the buildup of this charge. It follows from (4.15) and (4.18) that

$$x(t) = 2 \left[\frac{J_{pC}\tau_\nu D_n}{J_C} (1 - e^{-t/\tau_\nu}) \right]^{1/2} \tag{4.19}$$

The voltage drop across the unmodulated part of the ν-region is obtained by combining (4.11) and (4.19):

$$V(t) = \frac{J_C}{q\mu_n N_\nu} \left\{ W_\nu - 2 \left[\frac{J_{pC}\tau_\nu D_n}{J_C} (1 - e^{-t/\tau_\nu}) \right]^{1/2} \right\} \tag{4.20}$$

Neglecting the offset voltage due to ($V_{BE} - V_{BC}$), as well as the voltage drop in the modulated part of the ν-region, (4.20) describes the output voltage during turn-on, once the device enters the quasi-saturation region. In a typical power transistor, τ_ν is kept large to minimize quasi-saturation effects. Consequently the turn-on transient for these devices is correspondingly slow.

The turn-off process for a p^+-n-ν-n^+ transistor is also relatively complex, since all the stored charge in the ν-region must be removed before the device comes out of saturation. To an approximation, this is similar to a p^+-ν-n^+ diode with a reverse voltage applied across its terminals. Recovery proceeds by sweeping of holes out of the p-base and electrons out of the n-collector, with the flow of full reverse current. At $t = t_{OL}$ (see Section 3.3.2), a space charge region begins to build up across the p-ν junction, and the voltage across the device grows as this space charge layer expands. Thus, the change in collector-emitter voltage is almost entirely due to processes at the p-ν junction and has a three-halves power time dependence, as shown in (3.136a). However, an additional complication is that the process is two dimensional in nature, with a squeezing of the conduction region as described in Section 5.6.

4.3 CURRENT GAIN

The n^+-p-n transistor of Fig. 4.6 is biased into its active region (emitter-base diode forward biased, and collector-base diode reverse biased), and

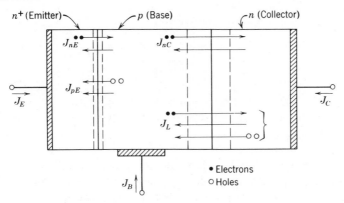

Fig. 4.6 Current components in an n^+-p-n transistor.

J_{nE} = electron component of emitter current density J_E injected into the base region

J_{pE} = hole-component of emitter current density, injected from the base into the emitter region

J_{nC} = electron component of collected current density J_C

Also given are the charge generation components of current due to deep levels in the collector-base depletion layer, and the diffusion-limited current terms. Together, these constitute the leakage current I_L. This term is not altered by changes in the emitter or base current and does not affect the ac properties of the transistor.

The common base current gain is defined as

$$\alpha = -\frac{\partial J_C}{\partial J_E} = \frac{-\partial J_{nE}}{\partial J_E} \frac{\partial J_{nC}}{\partial J_{nE}} \frac{\partial J_C}{\partial J_{nC}} \tag{4.21}$$

The first of these terms, $-\partial J_{nE}/\partial J_E$, is defined as the injection efficiency γ, $\partial J_{nC}/\partial J_{nE}$ is called the base transport factor α_T, and $\partial J_C/\partial J_{nC}$ is the collector multiplication factor M. The low frequency *common base current gain* is thus given by

$$\alpha_0 = \gamma \alpha_T M \cong \gamma \alpha_T \tag{4.22}$$

since the transistor is normally operated at collector-base voltages well below the breakdown value.

If the *low frequency common emitter current gain* is $\beta_0 = \alpha_0/(1-\alpha_0)$, and both γ and $\alpha_T \cong 1$, it can be shown that

$$\frac{1}{\beta_0} \cong \frac{1 - \alpha_T}{\alpha_T} + \frac{1 - \gamma}{\gamma} \cong \frac{1}{\gamma_e} + \frac{1}{\alpha_{Te}} \tag{4.23}$$

where γ_e and α_{Te} are the injection efficiency and base transport factors, respectively, for the common emitter connection, and are related to the common base values as shown.

The base transport factor is a measure of the electron current that crosses the base without recombination and is related to the base width W_B and the electron diffusion length L_{nB}. For transistors with uniformly doped base regions, it can be shown [1] that the base transport factor for the common emitter connection is given by

$$\alpha_{Te} \cong 2 \frac{L_{nB}^2}{W_B^2} \tag{4.24}$$

For example, as long as the base width is less than one-tenth the diffusion length, α_{Te} is greater than 200, and its effect on the current gain can be ignored.

The base transport factor plays a significant role in determining the current gain of high voltage n^+-p-n^+ transistors, in which extremely wide base widths (up to 150 μm in some devices) are encountered. On the other hand, the base width of n^+-p-ν-n^+ transistors is typically less than 10 μm; the current gain in these transistors is given almost entirely by the injection efficiency.

4.3.1 Injection Efficiency

The common emitter injection efficiency γ_e is given by

$$\gamma_e = \frac{J_{nE}}{J_{pE}} \tag{4.25}$$

Combining (3.62) and (3.63), we have

$$\gamma_e \cong \frac{D_n^* \int_{W_E}^{0} Ne^{-\Delta E_g/kT} e^{-x/L_{pE}}\, dx}{D_p^* \int_{0}^{W_B} N\, dx} \tag{4.26}$$

where D_n^* and D_p^* are the average values of electron and hole diffusion constants, N is the net ionized impurity concentration (donors minus acceptors), L_{pE} is the diffusion length of holes in the emitter, W_B is the base width, W_E is the emitter width, and ΔE_g is the reduction in energy gap due to high emitter doping effects. This equation may also be written more compactly in the form

$$\gamma_e = \frac{D_n^* Q_E}{D_p^* Q_B} \qquad (4.27)$$

where Q_E and Q_B are the effective charges due to ionized impurities in the emitter and base, respectively. Q_B increases with any increase in the base width or in the base doping level. The situation with Q_E is somewhat more complex; that is, Q_E increases as the emitter doping concentration is increased. However at doping levels above $1.85 \times 10^{19}/\text{cm}^3$, Q_E falls because of the narrowing of the energy gap in silicon (see Section 3.2.1). Additionally, heavy doping in the emitter results in reducing the hole lifetime, hence the hole diffusion length L_{pE}, further reducing Q_E. In typical diffused emitter devices, the value of Q_E changes with the emitter surface concentration, so that the current gain exhibits a broad maximum at around 10^{19} to $10^{20}/\text{cm}^3$ for this parameter.

The diffusion length of holes in the emitter L_{pE} is governed by the minority carrier lifetime. This topic was covered in Section 1.1 and is summarized here. Two recombination processes affect the minority carrier lifetime in a heavily doped emitter. The first is phonon-assisted indirect recombination, via deep impurity levels in the energy gap. For an n^+-type semiconductor, the time rate of change of excess minority carrier concentration due to this process is given by

$$-\frac{dp'}{dt} = \gamma_1 p' \qquad (4.28)$$

where $\gamma_1 = 1/\tau_p$.

A second process, direct band-to-band recombination with the transfer of excess energy to a mobile electron, also occurs in heavily doped n-emitters. This process, known as Auger recombination, is characterized (1.20a) by

$$-\frac{dp'}{dt} = \gamma_{3p} n^2 p' \qquad (4.29)$$

where $\gamma_{3p} = 1.7 \times 10^{-31}$ cm^6/sec for silicon and n is the electron concentration in the emitter, $= N_E^+$. By analogy with indirect recombination, the

Auger lifetime is given by

$$\tau_A = \frac{1}{\gamma_{3p} N_E^{+2}} \tag{4.30}$$

Finally, the effective lifetime in the emitter τ_{eff} is given (1.21d) by

$$\frac{1}{\tau_{eff}} = \frac{1}{\tau_p} + \frac{1}{\tau_A} \tag{4.31}$$

It is interesting to compare the magnitudes of these terms and their relative effects on transistor gain. Consider first a uniformly doped emitter, with an ionized impurity concentration of $2 \times 10^{19}/\text{cm}^3$. For this situation, $\tau_A \cong$ 14.7 nsec. This is comparable to the low level lifetime due to indirect recombination. Thus both processes are of equal significance in their effect on the gain. For higher doping levels, however, the Auger lifetime falls off more rapidly, so that the Auger process will dominate.

In a diffused transistor, the ionized impurity concentration increases exponentially as we move away from the depletion layer edge, toward the emitter surface. Consequently the comparison becomes more complex. Qualitatively, however, for impurity levels of interest, we note that τ_A varies inversely with the square of the ionized doping level, whereas τ_p varies approximately inversely with the doping level. Therefore the integration limit for L_{pE} (hence the magnitude of Q_E) in deep-diffused emitters is set by Auger recombination. With shallow emitters, however, the integration limit is determined by the width of the emitter region.

A rough estimate of the temperature dependence of the common emitter current gain can be made, based on the assumption that all the variation is due to band-gap narrowing effects. For a uniformly doped long emitter region, (4.26) can be reduced to the form

$$\beta_0(T) = \text{const} \times e^{-\Delta E_g / kT} \tag{4.32}$$

where β_0 is the low frequency common emitter current gain. Thus the current gain increases with temperature, and the rate of increase for heavy emitter doping is larger than for light doping. Combining (4.26), (4.28), and (4.30), it can be shown that β_0 increases by a factor of about 1.52 between 25 and 155°C if the emitter is uniformly doped at $5 \times 10^{19}/\text{cm}^3$, and by a factor of about 3.46 if the doping level is $2 \times 10^{20}/\text{cm}^3$. In a practical double diffused transistor, where the emitter is graded, β_0 typically increases [6] by a factor of 2 over this temperature range.

4.3.2 β-Falloff at High Levels: Emitter Effects

A number of effects on the emitter side lead to a reduction in current gain at high injection levels. First, with increasing emitter current, the minority carrier concentration in the base increases to the point where the concurrent change in majority carrier concentration becomes significant. Although a similar effect is present for the emitter region, this region is much more highly doped, and the conductivity modulation effect is insignificant. As a consequence, the injection efficiency of the transistor decreases with increasing forward current, leading to a falloff in the current gain characteristic. In addition, an $\mathscr{E}$ field is established in the base region because of the variation of carrier concentration with distance. This field aids the diffusion of minority carriers; in the limit it results in an effective doubling of the diffusion constant in the base region. Finally, the current distribution in the emitter becomes nonuniform at high injection levels. This crowding results in reinforcing the conductivity modulation effect in localized regions and produces a premature decrease in the current gain. These effects are now considered.

Under low level injection conditions, we can write

$$\gamma_{e,\text{low}} = \frac{D_n^*}{D_p^*} \frac{Q_E}{Q_B} \tag{4.33}$$

where Q_B, the net charge due to the impurity concentration in the base, is given by $\int_0^{W_B} N(x)\,dx$. With injection, an additional charge Q_S is stored in the base and is balanced by an equal and opposite majority carrier charge. The injection efficiency under conditions of current flow is given by

$$\gamma_e = \frac{\gamma_{e,\text{low}}}{1 + Q_S/Q_B} \tag{4.34a}$$

From charge control, the stored charge in the base is linearly proportional to the collector current. Thus the injection efficiency can be written in a more convenient form, as

$$\gamma_e = \frac{\gamma_{e,\text{low}}}{1 + J_C/J_0} \tag{4.34b}$$

and falls once the collector (hence, emitter) current density exceeds some critical value J_0.

A second effect is current crowding under the emitter. In device operation the majority carrier base current flows laterally as in Fig. 4.7a. The

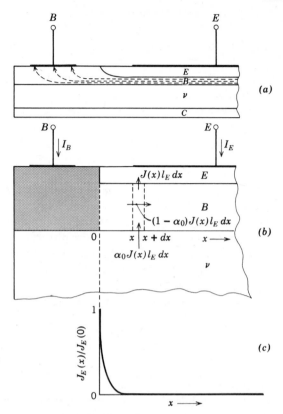

Fig. 4.7   Emitter current crowding.

flow of this majority carrier current results in a potential drop in the lateral direction, so that different regions of the emitter-base junction are biased to a different forward voltage. This, in turn, gives rise to uneven injection* over the width of the emitter, for a transistor geometry of the type shown. Here the majority of the injection occurs from the edge of the emitter nearest to the base contact. The use of a second base contact effectively doubles the injecting width of the emitter.

The degree of emitter crowding may be determined [7] by considering the idealized transistor of Fig. 4.7b. The following assumptions are made.

1. The emitter region represents an equipotential plane.
2. The entire base contact as well as the region between it and the edge of

─────────────────

*Note that a lateral ohmic drop of about 26 mV results in a reduction of the injected emitter current density by a factor of $1/e$.

the emitter (cross-hatched in Fig. 4.7b) is considered to be an equipotential region.
3. The current gain is constant for the device over the entire width of the emitter. Although not essential, this assumption greatly simplifies the mathematical analysis, while still preserving the physical aspects of the problem.
4. The emitter region is taken to have a length l_E and to be infinitely wide.

For high level injection conditions, the injected emitter current density at any point along the emitter width is given by

$$J(x) \cong J_0 e^{qV_{BE}(x)/2kT} \tag{4.35a}$$

so that

$$dJ(x) = \frac{q}{2kT} J(x) dV_{BE}(x) \tag{4.35b}$$

The base current at any point x is equal to the base current flowing up to this point. Thus

$$I_B(x) = I_B - \int_0^x (1 - \alpha_0) J(x) l_E dx \tag{4.36a}$$

and

$$dI_B(x) = -(1 - \alpha_0) J(x) l_E dx \tag{4.36b}$$

From (4.36b) it follows that the effective conductivity of the base region can be written as

$$\sigma_B' = \sigma_B \left[1 + \frac{J(x)}{J_0} \right] \cong \sigma_B J(x)/J_0 \tag{4.37}$$

where J_0 is the current density at which high level injection begins. Finally, from Ohm's law

$$dV_{BE}(x) = -\left[\frac{I_B(x)}{\sigma_B' W_B' l_E} \right] dx \tag{4.38}$$

where W_B' is an effective base width through which lateral current is assumed to flow.

Combining (4.35b), (4.37), and (4.38) gives

$$\frac{dJ(x)}{dx} = -\frac{q}{2kT}\frac{I_B(x)J_0}{\sigma_B W_B' l_E} \tag{4.39}$$

Differentiating, substituting into (4.36b), and noting that $I_E = I_B/(1-\alpha_0)$, we obtain

$$\frac{d^2 J_E(x)}{dx^2} - \frac{q}{2kT}\frac{J_0(1-\alpha_0)}{\sigma_B W_B'}J_E(x) = 0 \tag{4.40}$$

Solution of this equation results in an exponentially decaying emitter current density distribution that has a characteristic length x_0, where

$$x_0 = \left[\frac{2kT}{q}\frac{W_B'\sigma_B}{J_0(1-\alpha_0)}\right]^{1/2} \tag{4.41}$$

Thus it follows that injection from the emitter will be predominantly an edge effect. This causes a localized concentration in J_E over its average value, so that conductivity modulation of the base occurs prematurely.

It is interesting to note that once high level injection conditions are achieved, the effective width of the emitter injection region becomes independent of current density. This is a direct consequence of conductivity modulation of the base region described earlier. In high level operation, increasing collector current is accompanied by both increasing base current and increasing base conductivity. Consequently the lateral base voltage drop, hence the effective emitter injection width, becomes independent of further increase in current density at this point.

Current crowding effects in finite width emitters have also been determined [8], and are shown in Fig. 4.8 for an emitter of width w_E and a single base contact. Here emitter crowding is seen to be a function of the parameter ϕ, where

$$\phi = \frac{q}{2kT}\frac{w_E}{l_E}\frac{I_E(1-\alpha_0)}{W_B'\sigma_B} \tag{4.42}$$

Practical devices use base contacts on either side of the emitter for full area utilization. This situation can be considered as analogous to that of two single-contact devices in parallel. Thus Fig. 4.8 applies to this structure also if I_E, I_B, and w_E are replaced by $I_E/2$, $I_B/2$, and $w_E/2$, respectively.

The foregoing theory assumes that the temperature rise in the device is relatively small, making the emitter surface essentially isothermal. At high power dissipation levels, however, the center of the emitter stripe is at a higher temperature than its edge. Since the conduction characteristics of

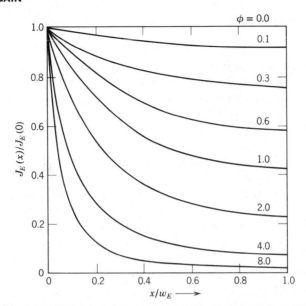

Fig. 4.8 Current distribution for a two-stripe transistor geometry. Copyright © 1964 by the Institute of Electrical and Electronics Engineers, Inc. Reprinted with permission from Hauser [8].

junction diodes improve with temperature, there will be an increase in current density towards the center of the emitter, partially offsetting the effects of edge crowding due to the lateral base drop. Computer-aided solutions of current distribution in emitters with finite thermal conductivity have shown [9] that the temperature at the center of a finger can rise as much as 25°C over that at the edge. This can result in changing the emitter current profile so that at high power levels, the center of the emitter actually conducts more strongly than its edge.

From (4.41), $x_0 \cong W_B' \, (2 \text{ to } 4\beta_0)^{1/2}$. To a very rough approximation, therefore, the width of the injecting edge is 10 to 14 W_B (assuming $\beta_0 \cong 50$), or about 100 to 140 μm for a typical power device having a base width of 10 μm. Consequently the emitter stripe can be as wide as 200–280 μm for its full utilization. This is quite easy to achieve by modern photolithographic methods.

Thus a power device that injects over its full emitter area* can be designed relatively easily by making the emitter thin and long. The shape of the overall structure is kept roughly square by using one or more emitter stripes alternating with base contact stripes. This is the well-known interdigitated structure in common use today.

---

*This does not mean, however, that the current density remains uniform with temperature or operating conditions.

4.3.3 β-Falloff at High Levels: Collector Effects

The current gain of a transistor is also affected by phenomena associated with the collector-base junction. These phenomena ultimately result in an expansion of the effective base width at high current densities and in the diffusion of minority carriers through this extended base region. Thus the charge stored in the base, Q_B, increases and the injection efficiency falls [see (4.27)]. This, in turn, causes a fall in the common emitter current gain at high injection levels.

A second effect of the increase in base width is a fall in the base transport factor α_{T_e}. This term is important in determining the falloff in current gain for wide base, high voltage, n^+-p-n^+ transistors. However it has little significance in the operation of narrow base n^+-p-v-n^+ devices.

The effect of increasing collector current in an n^+-p-n^+ transistor is straightforward and is considered for a device with a uniform base concentration N_B. Let W_B and x_{B0} be the base and depletion layer widths for zero collector current, respectively. When a forward current flows, resulting in the transport of electrons from the emitter to the collector, n' is the concentration of mobile electrons traversing the collector depletion layer. Thus the current density is given by

$$J_C = qn'v_{\lim} \tag{4.43}$$

where $v_{\lim}$ is the limiting drift velocity (10^7 cm/sec) at which electrons move through the depletion layer. The presence of mobile electrons in the depletion layer increases the space charge density to $N_B + n'$, so that its thickness shrinks to a new value x_B. Since the collector is heavily doped, the depeltion layer width in this region is unchanged. Solution of Poisson's equations for this situation gives

$$x_B = \frac{x_{B0}}{(1 + n'/N_B)^{1/2}} \tag{4.44}$$

assuming uniform doping. Setting $J_0 = qv_{\lim}N_B$, this can be rewritten in the form

$$x_B = \frac{x_{B0}}{(1 + J_C/J_0)^{1/2}} \tag{4.45}$$

where J_0 is a critical current density at which widening of the base width becomes significant. γ_e and α_{T_e} are reduced by this base-widening effect, resulting in a fall in the current gain.

The situation for an n^+-p-ν-n^+ transistor is considerably more complex, and a number of different cases can obtain [10, 11], depending on the bias conditions. As before, forward current flow results in the transport of electrons from the emitter to the collector, through the base-ν depletion layer. In the n^+-p-ν-n^+ transistor, however, the base doping is high relative to that of the ν-region; therefore the effect of mobile electrons is only significant in the ν-side. Here, increasing collector current results in lowering the space charge density to N_ν-n', where n' is the mobile carrier concentration. Thus the depletion layer *tends to expand*. At the same time, however, the increase of forward collector current results in an increase in the ohmic drop across the undepleted part of the ν-region, causing a reduction of the available voltage bias across the p-ν junction. Consequently the depletion layer *tends to shrink*. One of these effects will dominate, for any given situation. In both situations, however, it can be shown that the end result is an effective widening of the base width when the collector current density exceeds a specific critical value. Again, this is accompanied by an increase in Q_B and a fall in the common emitter current gain.

Consider an n^+-p-ν-n^+ transistor, fed from a constant collector supply voltage. Assume that its depletion layer extends x_0 into the ν-region for $J_C = 0$ (i.e., the ν-region is only partially depleted). With increasing collector current, the ohmic drop across the undepleted part of the ν-region increases, until all the supply voltage is used up across this region and the depletion layer collapses. If J_0' is the critical current at which this occurs, then from Ohm's law we have

$$J_0' = \frac{q\mu_\nu N_\nu (V_{CB} + \psi)}{W_\nu} \qquad (4.46)$$

where μ_ν, N_ν, and W_ν are the electron mobility, the ionized carrier density, and the width of the ν-region, respectively, and ψ is the contact potential of the p-ν junction.

This situation is commonly encountered with low supply voltages. Here the $\mathcal{E}$ field in the ν-region is given by $(V_{CB} + \psi)/W$. Typically this is a few hundred volts per centimeter; that is, it represents *the low field case*. Transistor action at $J_C = J_0'$ consists of diffusion through W_B, drift through a low field ν-region of width W_ν, and eventual collection.

The p-ν junction becomes forward biased when the current density exceeds J_0'. This is accompanied by hole injection into the ν-region, together with conductivity modulation of part of it, as in Fig. 4.4b. Thus for $J_C > J_0'$, the drift region shrinks *away* from the p-ν junction, reducing its width from W_ν to $(W_\nu - W_{CIB})$. For this situation, J_C is obtained by modifying (4.46) so that

$$J_C = \frac{q\mu_\nu N_\nu (V_{CB} + \psi)}{W_\nu - W_{CIB}} \qquad (4.47)$$

where W_{CIB} is the current-induced base width—that is, the modulated part of the ν-region, $0x'$ in Fig. 4.4b. Combining (4.46) with (4.47) gives its value as

$$W_{CIB} = W_\nu \left(1 - \frac{J_0'}{J_C} \right) \qquad (4.48)$$

Transistor action at a current density of $J_C > J_0'$ is thus one of diffusion through a neutral region of width $(W_B + W_{CIB})$, drift through a region of width $(W_\nu - W_{CIB})$, and eventual collection. Again, the expansion of the base width results in an increase of Q_B and a fall in γ_e.

Figure 4.9 gives the output characteristic of an n^+-p-ν-n^+ transistor. The device operation just described occurs when the operating point moves along the path ABC, for a base current of I_{B1}. Here collapse of the p-ν depletion layer occurs at point B; further increase of collector current density moves the operating point into the quasi-saturation region BC.

Expansion of the depletion layer usually occurs for high values of supply voltage. As before, assume a depletion layer width of x_0 for $J_C = 0$. If n' is the electron concentration entering the depletion layer, then

$$x = \frac{x_0}{(1 - n'/N_\nu)^{1/2}} \qquad (4.49)$$

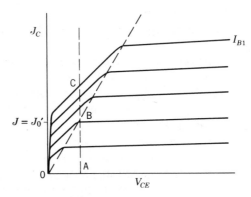

Fig. 4.9   High level operation.

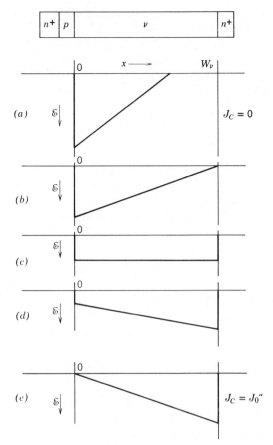

Fig. 4.10 Electric field profiles at high current levels.

where x is the width of this depletion layer, assuming uniform doping. Note that x expands with increasing n'.

Figure 4.10 shows the electric field configurations in the ν-region with increasing current density and for a constant collector-base voltage.* Initially, as in Fig. 4.10a, the depletion layer extends only partly into the ν-region but expands with increasing collector current density until it reaches to the ν-n^+ junction (Fig. 4.10b). With further increase in J_C, the $\mathscr{E}$ field profile changes, its peak value shifting from the p-ν junction to the ν-n^+ junction. Eventually we arrive at the situation of Fig. 4.10e, where the $\mathscr{E}$ field is zero at the p-ν junction and increases until it peaks at the

---

*As a result, all areas under these curves are equal.

ν-n^+ junction. Electrons enter this region at the p-ν junction. These are the essential conditions for *single injection*, and space charge limited current flow, as described in Section 1.3. Let $J_C = J_0''$ for this situation.

The value of the critical current density J_0'' can be calculated as follows. Writing $n(x)$ as the electron concentration in the ν-region, the space charge density is given by

$$\rho(x) = q\left[N_\nu - n(x) \right] \tag{4.50}$$

But

$$J_C = qv_d(x)n(x) \tag{4.51}$$

Substituting into (4.50) gives Poisson's equation as

$$\frac{d\mathcal{E}}{dx} = \frac{q}{\varepsilon\varepsilon_0}\left[N_\nu - \frac{J_C}{qv_d(x)} \right] \tag{4.52}$$

For high voltage operation we can make the approximation that $v_d(x) = v_{\text{lim}}$ and is constant over the entire ν-region. In addition, the contact potential can be ignored. Making these substitutions, (4.52) can be solved to obtain

$$\mathcal{E}(x) = \mathcal{E}(0) + \frac{qx}{\varepsilon\varepsilon_0}\left[N_\nu - \frac{J_C}{qv_{\text{lim}}} \right] \tag{4.53}$$

For the situation of Fig. 4.9e, the boundary values are

$$\mathcal{E}(0) = 0 \tag{4.54}$$

$$\mathcal{E}(W_\nu) = -2V_{CB}/W_\nu \tag{4.55}$$

since the voltage drop across the ν-region is V_{CB}. Combining with (4.53) gives

$$J_0'' = qv_{\text{lim}}\left(N_\nu + \frac{2\varepsilon\varepsilon_0 V_{CB}}{qW_\nu^2} \right) \tag{4.56}$$

This is the critical current density for operation in the regime where V_{CB}/W_ν is typically 10^4–10^5 V/cm, that is, *the high field case*. Consequently the approximation that $v_d(x) = v_{\text{lim}}$ is justified over most of the ν-region.

The critical collector current at which falloff in gain occurs is $J_0'' A_E$, where A_E is the area of the emitter. This corresponds to what happens when space charge limited conditions prevail in the ν-region. From (1.34) it is seen that the space charge limited current density is inversely proportional to the cube of the length of the space charge region. Consequently, a further increase in collector current can come about by a reduction in the effective length of W_ν. Alternately, collector current can increase if the area of the conduction path increases. Thus two different physical models can be proposed, and actual device operation usually can be explained by assuming that both models are operative in practice.

4.3.3.1 The Two-Dimensional Model

It has been proposed [12] for the two-dimensional case that the cross-sectional area to current flow in the ν-region expands, so that the density remains at J_0'' in this region, with increasing J_C. With reference to Fig. 4.11, if w_E is the width of the emitter (uniform injection is assumed here) and w is the width of the current flow path in the ν-region, we have

$$\frac{w}{w_E} = \frac{I_C}{J_0'' A_E} \tag{4.57}$$

The effective base width thus expands from W_B to a new value, $W_B' \cong AB$.

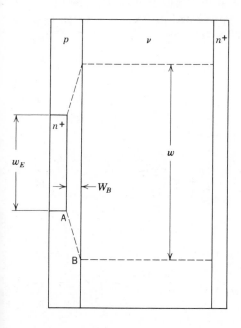

Fig. 4.11 Current beaming effects.

From geometrical considerations

$$W'_B = \left[W_B^2 + \frac{(w - w_E)^2}{4} \right]^{1/2} \tag{4.58}$$

$$= W_B \left[1 + \frac{w_E^2}{4 W_B^2} \left(\frac{I_C}{J_0'' A_E} - 1 \right)^2 \right]^{1/2} \tag{4.59}$$

Note that the increase in base width, which occurs when I_C exceeds $J_0'' A_E$, is current induced.

4.3.3.2 The One-Dimensional Model

Figure 4.10e corresponds to the situation where all the carriers are moving with a velocity $v_{\lim}$ at the onset of saturation. Since ρ is constant in the space charge layer, the field distribution is linear, as shown. An increase in J_C above J_0'' must be accompanied [13] by an increase in injected carrier density, thus by an increase in the electron space charge density. This in turn results in an increase in the slope $d\mathscr{E}/dx$. However $\int \mathscr{E}\, dx$ is constant and has a magnitude equal to V_{CB}. Thus the distance over which the $\mathscr{E}$ field exists must decrease, resulting in the formation of a current-induced base region W_{CIB}. This process is represented in Fig. 4.12, which extends the process depicted in Fig. 4.10. With reference to Fig. 4.12b,

$$V_{CB} = - \left[\frac{(W_\nu - W_{CIB})^2}{2} \right] \frac{d\mathscr{E}}{dx} \tag{4.60}$$

Substituting into (4.52) gives

$$V_{CB} = \frac{(W_\nu - W_{CIB})^2 q}{2\varepsilon\varepsilon_0} \left(N_\nu - \frac{J_C}{q v_{\lim}} \right) \tag{4.61}$$

At $J = J_0''$, $W_{CIB} = 0$, so that

$$V_{CB} = - \frac{W_\nu^2 q}{2\varepsilon\varepsilon_0} \left(N_\nu - \frac{J_0''}{q v_{\lim}} \right) \tag{4.62}$$

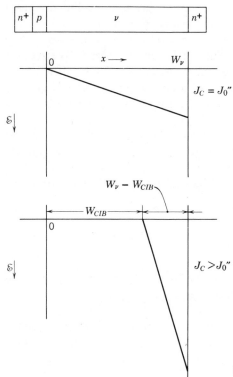

Fig. 4.12 Formation of the current-in-duced base.

Combining (4.61) with (4.62),

$$W_{CIB} = W_\nu \left[1 - \left(\frac{J_0'' - qv_{\lim}N_\nu}{J_C - qv_{\lim}N_\nu} \right)^{1/2} \right] \qquad (4.63)$$

Thus following this model W_{CIB} increases with increasing J_C, and the effective base width expands once J_0'' is exceeded.

In summary, two distinct possibilities are present when the current density in an n^+-p-ν-n^+ transistor is increased. First, the increasing voltage drop across the ν-region reduces the available reverse bias at the collector-base junction. Consequently its depletion layer *shrinks* and eventually collapses. Further increase of collector current drives the device into quasi-saturation, accompanied by an expansion of the base width. This process is commonly encountered for low supply voltages and is known as the low field case.

Conversely, increasing collector current results in reducing the space charge density in the p-ν depletion layer, causing it to *expand* until it fills the ν-region. Further increase in collector current results in space charge limited current flow in this region. Still further increase leads to growth in the effective base width, by one or more mechanisms, described by the one- and two-dimensional models. These processes are commonly encountered with higher supply voltages, which means that the device does not enter quasi-saturation. Typically, $N_\nu W_\nu = 10^{11}/\text{cm}^2$ represents the boundary between the low field case and the high field case. In both situations, increase of the effective base width results in increasing the value of Q_B, resulting in a drop in the injection efficiency with increasing collector current density. This is reflected in the falloff of current gain at high levels.

4.4   GAIN–BANDWIDTH PRODUCT

A simple expression for the gain-bandwidth product of a transistor, derived from charge control principles [1], is

$$\frac{1}{2\pi f_t} = \frac{kT}{qI_C}C_{tE} + t_B + r_C C_{tC} + \frac{x_C}{v_{\text{lim}}} \tag{4.64}$$

where f_t is the gain-bandwidth product, C_{tE} and C_{tC} are the average emitter and collector transition capacitances, x_C is the width of the collector-base depletion layer, r_C is the collector body resistance, and t_B is the base transit time. Typically $t_B = W_B^2/4D_{nB}$, since our interest is in high level operation.

A number of modifications must be made before this equation can be used to describe the behavior of power devices at high current levels.

1. The term involving C_{tE} can be ignored at high current levels.
2. The term involving C_{tC} can be ignored, since the collector body resistance is extremely small.
3. The transit time through the depletion layer can also be ignored, since carriers move through it at the limiting velocity. The transit time through any high field space charge region can be ignored, for the same reason.
4. High level injection results in an effective widening of the base width, regardless of the actual mode of operation. This current-induced base must be included in determining the base transit time.

5. Finally, the transit time through the drift regions must be included if carrier velocity is low. This is especially true in the low field case, described earlier.

We now consider the individual situations that can arise.

4.4.1 The n^+-p-n^+ Transistor

The true base width of the n^+-p-n^+ transistor increases by $W_{CIB} = x_{B0} - x_B$ as given by (4.45). Thus the gain-bandwidth product is given from (3.25) by

$$\frac{1}{2\pi f_t} \cong \frac{(W_B + W_{CIB})^2}{4D_{nB}} \tag{4.65}$$

for high level injection.

4.4.2 The n^+-p-ν-n^+ Transistor: Low Field Case

Here, in addition to the current-induced base, carriers move by drift through the remainder of the ν-region. The gain bandwidth product is thus given by

$$\frac{1}{2\pi f_t} \cong \frac{W_B^2}{4D_{nB}} + \frac{W_{CIB}^2}{4D_\nu} + \frac{(W_\nu - W_{CIB})^2}{\mu_\nu(V_{CB} + \psi)} \tag{4.66}$$

since the entire potential drop $V_{CB} + \psi$ is supported by $W_\nu - W_{CIB}$. Note that μ_ν and D_ν are the low field mobility and diffusion constants, respectively, for electrons in the ν-region. The value of W_{CIB} is given by (4.48) as

$$W_{CIB} = W_\nu \left(1 - \frac{J_0'}{J_C}\right) \tag{4.67}$$

where J_0' is the current density at which the p-ν junction becomes forward biased.

4.4.3 The n^+-p-ν-n^+ Transistor: High Field Case

There are two models to describe the base-widening effect in the n^+-p-ν-n^+ transistor in the high field case. Invoking the two-dimensional model, carriers diffuse through a base whose effective width is given by

$$W_B' = W_B \left[1 + \frac{w_E^2}{4 W_B^2} \left(\frac{I_C}{A_E J_0''} - 1 \right)^2 \right]^{1/2} \tag{4.68}$$

where J_0'' is the current density at the onset of space charge limited current flow. In addition, they drift through the ν-region at their limiting velocity, permitting us to ignore the transit time through this region. Thus the gain-bandwidth product can be written as

$$\frac{1}{2\pi f_t} \cong \frac{W_B^2}{4 D_{nB}} \left[1 + \frac{w_E^2}{4 W_B^2} \left(\frac{I_C}{A_E J_0''} - 1 \right)^2 \right] \tag{4.69}$$

For the one-dimensional model,

$$\frac{1}{2\pi f_t} \cong \frac{W_B^2}{4 D_{nB}} + \frac{W_{CIB}^2}{4 D_\nu} \tag{4.70}$$

where W_{CIB} is given by (4.63). Again, the transit time through the remainder of the ν-region can be ignored because of the prevailing high field conditions.

4.5 INSTABILITIES AND SECOND BREAKDOWN

A number of different types of electrical instability are exhibited by transistors, depending on the mode of circuit operation. This is to be expected; the transistor is an active device, capable of substantial power gain, and it can be made unstable if external or internal conditions provide the necessary positive feedback.

In addition to electrical instabilities, the transistor is also capable of exhibiting pure thermal instability. This comes about because the conduction characteristics of the emitter-base diode improve with temperature, while the thermal conductivity of silicon gets worse (see Section 1.5.2). Furthermore, since the current gain of a transistor increases with temperature, electrical and thermal instabilities generally act cooperatively within the device.

Electrical instabilities of the CCNR type lead to filamentary conduction and the formation of microplasmas, as described in Section 1.4.3. If allowed to persist so that the local temperature exceeds the intrinsic temperature for the material, this results in thermal instability and second breakdown, by way of the formation of a mesoplasma. Needless to say, the thermal instability by itself presents an alternate route to second breakdown, without the preliminary step of microplasma formation.

4.5.1 Lateral Electrical Instability

A transistor is particularly susceptible to second breakdown if it is operated under conditions of reverse base current flow [14]. This situation is encountered during turn-off, often resulting in a growing lateral electrical instability, which is initiated by inhomogeneous current flow in the base region.

The transistor in Fig. 4.13 is operating with terminal currents as shown, and the emitter current density initially is assumed to be uniform. Now consider the situation at a point X in the base, through which additional current flows, because of an inhomogeneity in one or more regions of the transistor. In any event, the electron and hole concentrations at this point are now in excess of the values in the neighboring material.

Under these conditions, the excess electrons are collected by normal transistor action. The holes move toward the base contact to provide the base current, setting up a potential drop in the base region, as shown. Thus the emitter-base diode becomes more forward biased at X with respect to the emitter edge, ensuring that more forward current flows at this point,

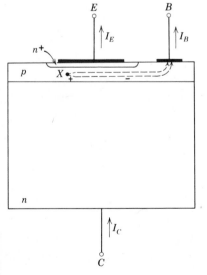

Fig. 4.13 Current pinching effects.

leading to a further localized increase in the emitter current, and so on. Eventually, all the emitter current pinches into this region, resulting in the formation of a mesoplasma.

In normal operation, base current flows in an opposite direction to that shown in Fig. 4.13. Consequently a lateral electrical instability of this type can occur only if the hole current created by the inhomogeneity is large enough to allow the direction of base current flow to be reversed. Thus this type of instability can be initiated at low current levels if reverse base current flows, but only at much higher current levels in normal device operation.

Local avalanche multiplication due to inhomogeneities in the collector-base depletion layer has been proposed by many workers [15, 16] as a mechanism for initiating this lateral electrical instability. Possibilities for this mechanism are present at oxygen inclusions, crystal defects, metallic impurities, deep traps, and at points of high depletion layer curvature. In all cases the generated electrons are removed by transistor action in the emitter-collector circuit, whereas the holes leave by way of the base and initiate this form of lateral instability.

Electrical inhomogeneity can also result from filamentary conduction caused by CCNR behavior, as described in Section 1.4.3. This represents an important alternative initiating mechanism to the defect-induced inhomogeneities described in the last paragraph. This can occur in two situations that have special importance because they do not depend on materials or process-induced inhomogeneities. Thus second breakdown can occur in even "perfectly" processed devices, made from ideally uniform starting materials.

4.5.1.1 Device Operation at High Voltages

The output characteristics of a transistor can be shown to exhibit CCNR behavior at voltages in excess of the floating-base common emitter breakdown voltage, referred to as BV_{CEO}. The operation of a transistor with an inductive load is often accompanied by excursion of the operating point into this region of the device characteristics.

The common emitter current gain of a transistor is obtained from (4.22) as

$$\beta_0 = \frac{\gamma \alpha_T M}{1 - \gamma \alpha_T M} \tag{4.71}$$

For junction transistors, both γ and α_T are very close to, but less than, unity. The multiplication factor, however, exceeds unity with increasing collector-base voltage. Thus from (2.17) and (2.18),

$$M = \frac{1}{1 - (V/BV_{CBO})^m} \qquad (4.72)$$

where V is the applied voltage, BV_{CBO} is the breakdown voltage of the collector-base junction (emitter open), and $m = 4$ for n-p-n transistors and 6 for p-n-p transistors. With increasing voltage, β_0 eventually becomes infinite when $M = 1/\gamma\alpha_T$, and the transistor exhibits "breakover" characteristics. The voltage at which this occurs in BV_{CEO} and its value is obtained by combining (4.71) and (4.72). Thus

$$BV_{CEO} = \frac{BV_{CBO}}{(\beta_0)^{1/m}} \qquad (4.73)$$

This is illustrated in the common emitter characteristics (Fig. 4.14) that break over at BV_{CEO}. At normal forward currents, $BV_{CEO} \ll BV_{CBO}$, since β_0 is reasonably large ($\cong 50$–100). At extremely low emitter current levels, however, β_0 falls, and the breakdown voltage rises to its upper limit of

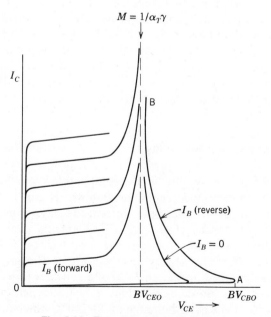

Fig. 4.14 Transistor characteristics.

BV_{CBO}. No sooner is this voltage exceeded, however, than the current through the device increases, accompanied by an increase in the current gain. Consequently the characteristic snaps back as shown (0AB), resulting in a CCNR characteristic. Operation in this region leads to filamentary conduction, which can initiate the lateral electrical instability. The time for initiation of this avalanche-induced negative resistance is extremely short, and is essentially determined by the circuit in which the device is imbedded.

4.5.1.2 Device Operation at High Currents

We have demonstrated that an electron space charge forms in the ν-region of a high voltage n^+-p-ν-n^+ transistor at high current densities. With increasing current, the $\mathscr{E}$ field configuration in this region shifts so that it peaks at the ν-n^+ junction. Eventually one carrier space charge limited current flow conditions are established at some value of critical current density. Further increase of J_C results in the formation of a current-induced base region and an increase in the peak $\mathscr{E}$ field at the ν-n^+ junction. Eventually this process [17] leads to the avalanche generation of hole-electron pairs when the peak electric field intensity exceeds 1.2×10^5 V/cm. The holes flow toward the p-ν contact, causing current conduction to shift from single injection to double injection. This process, which is essentially one of avalanche injection, results in the formation of a CCNR characteristic, as described in Section 1.4.1. Again, current conduction is filamentary and can initiate a lateral electrical instability in the base. Avalanche injection in the ν-region is also an extremely fast process, and its initiation time is dictated by circuit characteristics. The process of turning off an inductively loaded transistor by a reverse base bias often results [18] in second breakdown, which is initiated by this mechanism.

4.5.2 Lateral Thermal Instability

The lateral thermal instability is a second mechanism that can lead to mesoplasma formation and second breakdown. It can be initiated by a nonuniformity in the emitter-base current density, caused by a local inhomogeneity, by emitter crowding, or by filamentary current flow. This type of instability is likely to occur when the transistor is forward biased and grows because of the negative temperature coefficient of the emitter-base diode, as described in Section 3.4. For a constant forward current, the voltage across a junction diode falls linearly with temperature, at a rate of about 2 mV/°C. Thus over the range of 30–200°C, the emitter-base diode can be characterized by (3.142)

$$I_E = I^* \exp\left[\frac{q(V_{BE} + \phi \Delta T)}{kT}\right]$$ (4.74)

where $\phi = 2$ mV/°C, and I^* is a temperature-independent constant for a given diode.

The criteria for this instability can be evaluated [19] by considering a transistor to be made of a number of small identical devices in parallel. Two such devices appear in the circuit of Fig. 4.15a, where R and R' represent either internal parasitics or externally added resistors. The two devices are fed in parallel from a constant current base drive I, and currents I_1, I_2 flow in their respective collectors. The transistors are assumed to have high current gain so that $I_E \cong I_C$.

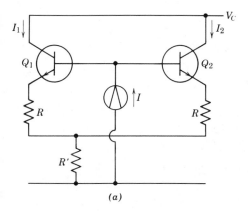

(a)

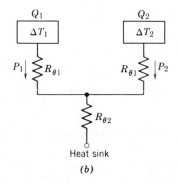

Heat sink
(b)

Fig. 4.15 Thermally coupled transistors.

The thermal circuit for this two transistor analog is given in Fig. 4.15b: device dissipations are P_1 and P_2, corresponding to temperature increments of ΔT_1 and ΔT_2, respectively. Thermal impedances associated with each device are $R_{\theta 1}$, with a common thermal impedance $R_{\theta 2}$ as shown. Finally, it is assumed that nearly all the power dissipation is at the collector-base junctions.

From (4.74), we can write

$$V_{BE1} = \frac{kT}{q} \ln \frac{I_1}{I_1^*} - \phi \Delta T_1 \tag{4.75a}$$

$$V_{BE2} = \frac{kT}{q} \ln \frac{I_2}{I_2^*} - \phi \Delta T_2 \tag{4.75b}$$

Also,

$$\Delta T_1 = (R_{\theta 1} + R_{\theta 2})P_1 + R_{\theta 2}P_2 \tag{4.76a}$$

$$\Delta T_2 = (R_{\theta 1} + R_{\theta 2})P_2 + R_{\theta 2}P_1 \tag{4.76b}$$

and

$$P_1 \cong I_1 V_C \tag{4.77a}$$

$$P_2 \cong I_2 V_C \tag{4.77b}$$

The voltage across the current source is given by

$$V_{BE1} + I_1 R + (I_1 + I_2)R' = V_{BE2} + I_2 R + (I_1 + I_2)R' \tag{4.78}$$

Combining these equations gives

$$I_2 = \frac{\dfrac{kT}{q} \ln\left(\dfrac{I_1}{I_2}\right)}{\left(\dfrac{I_1}{I_2} - 1\right)(R - R_{\theta 1} \phi V_C)} \tag{4.79}$$

Thus the current division is a function of the individual currents, as well as of the electrical and thermal parameters.

A number of important conclusions can be derived from a study of this equation. For example,

1. Lateral thermal stability is not dependent on the shared electrical or thermal resistances (R' or $R_{\theta 2}$) but only on the unshared terms (R and $R_{\theta 1}$).
2. The circuit can be stabilized by the inclusion of an appropriate value of a resistive element R in each separate emitter.
3. The circuit is stable to higher currents as R is increased or $R_{\theta 1}$ decreased.
4. For $R \geqslant R_{\theta 1} \phi V_C$, the circuit is stable for all values of current.

4.5.3 Corrective Measures

A number of corrective measures can be used to improve the second breakdown characteristics of transistors. Essentially, they consist of eliminating or minimizing the mechanisms by which electrical and thermal instabilities are initiated in a device. Techniques include the use of processes that result in a minimum of crystal damage, metallic impurities, doping inhomogeneities, or deep levels. These are described in Chapter 6. Another technique involves reduction of the thermal resistance of the device. This can be done by the lapping down the silicon slice to a minimum thickness that can be handled without breakage, or by adding another heat removal path by making thermal contact to both faces of the silicon die. Proper device and circuit design should ensure that under operation, internal electric fields cannot become excessively large and cause avalanche injection at the ν-n^+ contact. Finally, appropriate measures must be taken to confine circuit operation to voltage below BV_{CEO}.

A powerful technique for reducing lateral electrical and thermal instability is to break up the transistor into many small devices, as described in Fig. 4.15. This is readily done by subdividing the emitter region of a large device into many units, and isolating them by placing small resistors R in each emitter lead. Such a technique is known as emitter ballasting. Ballast resistors can also be placed in the base leads of a transistor. As a rule, however, these must be quite large to be effective, and they result in deterioration of the electrical properties of the device.

One approach to emitter ballasting is to design the transistor with a large number of separate emitters [20]. Figure 4.16 gives the V-I characteristics of one such emitter-base diode, for specified limits of allowable operating temperature T_{hot} and T_{cold}. Also shown are the design limits for its forward current, as I_{hot} and I_{cold}, respectively. The slope of the load line through these points gives the effective value of ballast resistor to be placed in each emitter lead. A more refined approach [21], based on the selection of a ballast resistor to maintain stable operation for a range of operating currents, can also be used here.

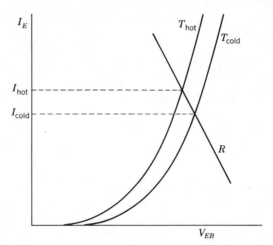

Fig. 4.16   Emitter-base characteristics.

Significant improvement in the power handling capability of a transistor can be achieved if the emitter is finely subdivided. Furthermore, the problem of making connections to individual emitters can be avoided by using a distributed emitter resistance. This takes the form [22] of a resistive film such as nichrome, which is evaporated over the emitter periphery, as in Fig. 4.17. Improvement in power handling capability by a factor of $4-8$ have been readily obtained by this technique.

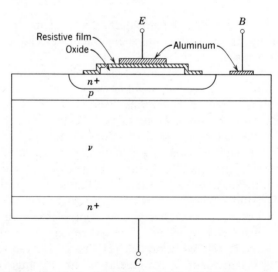

Fig. 4.17   Resistive metal film ballasting.

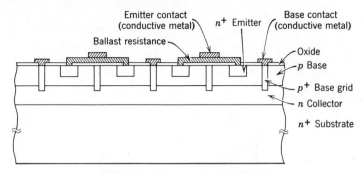

Fig. 4.18   The overlay transistor.

The overlay transistor [23] is an excellent example of the practical implementation of distributed emitter ballasting in high frequency power transistors. Here the emitter is subdivided into a series of small units, and each emitter is provided with a wraparound base contact in the form of a deep-diffused region. Ohmic connection to the emitters is made two rows at a time, with both a resistive film and a conductive film. The base contact is directly made to alternate rows of the base grid, by means of a conductive film. In this manner, base and emitter terminals can be brought out on opposite sides of a silicon die without the necessity of crossovers.

A cross section of an overlay transistor is shown schematically in Fig. 4.18. The device consists of an n^+-substrate that serves as the collector contact. An n-collector is epitaxially grown on the substrate, followed by a diffused p-base. A deep-diffused p^+-contact mesh that reaches through to the n-collector is also provided. The emitter is subdivided into many small sections, one in each rectangle formed by the p^+-contact mesh. Resistive metal stripes are contacted to the emitters, two rows at a time. Aluminum contacts are made to these resistive stripes and brought out to an emitter contact pad. Simultaneously, an aluminum contact is made to the (alternate) exposed base rows and brought out to a base contact. Devices of this type have been made with as many as 512 emitters and are capable of operation in the gigahertz frequency range, with power dissipation capability in excess of 100 W at 25°C.

4.5.4   Safe Operation Area

An important item of information for the power circuit designer is the locus of $I_C - V_{CE}$, which marks the boundary between stable and unstable operation. This locus is shown in Fig. 4.19 for steady state operation and defines a safe operating area (SOA) for circuit design. Logarithmic scales

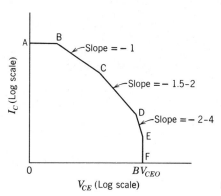

Fig. 4.19   The safe operation area.

are used for convenience. Features of interest in this SOA curve are as follows:

1. The line A-B represents the upper limit of current carrying capability. Operation at higher current levels results in failure by metallurgical damage to the bond leads, to the bond itself, or to the glass-to-metal seals on the header.
2. The region B-C indicates the basic thermal limitation associated with the maximum allowable junction temperature rise. As such, it is a constant power dissipation curve, and has a slope of -1.
3. The region C-D corresponds to second breakdown. It begins at relatively high current and low voltage and has been attributed [24] to high field effects at the ν-n^+ contact. Typically, it exhibits a slope of -1.5 to -2.
4. The region D-E is also associated with second breakdown, and it typically has a slope of -2 to -4. It is of relatively short extent and is often not observed on the SOA curve. It begins at low currents and high voltages and has been attributed [24] to effects at the p-ν junction.
5. The BV_{CEO} limit of the transistor is represented by EF.

The sections of the SOA curve labeled A, B, and C are based on considerations of steady state power dissipation. Thus for pulse operation, the SOA curve can be modified in this region, to encompass a larger area. The rest of the curve CDEF relates to second breakdown, which is a highly localized failure mode, relatively independent of the duty cycle.

4.5.5   Concluding Comments

The understanding of second breakdown phenomena in transistors is far from complete, and there is much conflicting opinion in this area. This is because experimental data are hard to obtain, given the destructive nature of this mechanism. Furthermore, the data are often difficult to evaluate when obtained, because of the large spread in the results. In general, however, it has been agreed that second breakdown is a destructive failure mode caused by mesoplasma formation at a highly localized site in the device. The initiation of this mesoplasma may come about by purely thermal means, or it may be preceded by current filamentation and the formation of a microplasma. It has been confirmed by many workers that high voltage operation is more conducive to second breakdown than operation at low voltages. Furthermore, operation under conditions of reverse base drive leads to earlier breakdown than for forward bias conditions.

There is also reasonable agreement that the time taken to initiate second breakdown is related to the applied power, as described in Section 1.5.2.2. Thus the $P\tau_D$ product is constant when the initiation time is short compared to the thermal relaxation time. For longer time delays, on the other hand, a $P\sqrt{\tau_D}$ = constant law is found to be more appropriate.

Some discrepancy has been noted [25] between the intrinsic temperature, as defined in Section 1.5.1, and the critical temperature as calculated on the basis of second breakdown data. Typically, critical temperatures have been found to be lower than the intrinsic by as much as 30°C in lightly doped collectors (5×10^{13}), and by even larger values in more heavily doped structures.

A number of workers [26–29] have attempted to characterize second breakdown phenomena by means of feedback models. Interest has been focused on the behavior of the emitter-base voltage of a transistor in its circuit environment. For a constant forward current, the emitter-base voltage falls at the rate of 2 mV/°C, and one criterion for stable operation is based on this property. Thus stable operation is maintained in any circuit in which the emitter-base voltage falls at a rate of 2 mV/°C or faster. On the other hand, a fall in the emitter-base voltage at a slower rate with increasing temperature is indicative of thermal runaway. Use of this approach has resulted in confirming the basic characteristics of second breakdown, and even in establishing the V_{CE}-I_C law in this region.

4.6 REFERENCES

1. S. K. Ghandhi, *The Theory and Practice of Microelectronics*, John Wiley & Sons, New York, 1968.

2. R. Denning and D. A. Moe, "Epitaxial π-ν n-p-n High Voltage Power Transistors," *IEEE Trans. Electron Devices*, **ED-17**, No. 9, pp. 711–716 (1970).

3. W. J. Chudobiak, "The Saturation Characteristics of npvn Power Transistors," *IEEE Trans. Electron Devices*, **ED-17**, No. 10, pp. 843–852 (1970).

4. L. E. Clark, "Characteristics of Two-Region Saturation Phenomena," *IEEE Trans. Electron Devices*, **ED-16**, No. 1, pp. 113–116 (1969).

5. P. L. Hower, "Application of a Charge-Control Model to High-Voltage Power Transistors," *IEEE Trans. Electron Devices*, **ED-23**, No. 8, pp. 863–870 (1976).

6. D. Buhanan, "Investigation of Current-Gain Temperature Dependence in Silicon Transistors," *IEEE Trans. Electron Devices*, **ED-16**, No. 1, pp. 117–124 (1969).

7. N. H. Fletcher, "Some Aspects of the Design of Power Transistors," *Proc. IRE*, **43**, No. 5, pp. 551–559 (1955).

8. R. J. Hauser, "The Effects of Distributed Base Potential on Emitter-Current Injection Density and Effective Base Resistance for Stripe Transistor Geometries," *IEEE Trans. Electron Devices*, **ED-11**, No. 5, pp. 238–242 (1964).

9. D. Navon and R. E. Lee, "Effect of Non-Uniform Emitter Current Distribution on Power Transistor Stability," *Solid State Electron.*, **13**, No. 7, pp. 981–991 (1970).

10. G. Rey and J. P. Bailbe, "Some Aspects of Current Gain Variations in Bipolar Transistors," *Solid State Electron.*, **17**, No. 10, pp. 1045–1058 (1974).

11. R. J. Whittier and D. A. Tremere, "Current Gain and Cut-Off Frequency Fall-Off at High Currents," *IEEE Trans. Electron Devices*, **ED-16**, No. 1, pp. 39–57 (1969).

12. A. van der Ziel and D. Agouridis, "The Cutoff Frequency Falloff in UHF Transistors at High Currents," *Proc. IEEE*, **54**, No. 3, pp. 411–412 (1966).

13. C. T. Kirk, Jr., "A Theory of Transistor Cutoff Frequency Falloff at High Current Densities," *IEEE Trans. Electron Devices*, **ED-9**, No. 2, pp. 164–174 (1966).

14. C. G. Thornton and C. D. Simmons, "A New High Current Mode of Transistor Operation," *IRE Trans. Electron Devices*, **ED-5**, No. 1, pp. 6–10 (1958).

15. H. A. Schafft et al., "Second Breakdown and Crystallographic Defects in Transistors," *IEEE Trans. Electron Devices*, **ED-13**, No. 11, pp. 738–742 (1966).

16. H. A. Schafft, "Second Breakdown—A Comprehensive Review," *Proc. IRE*, **55**, No. 8, pp. 1272–1288 (1967).

17. P. L. Hower and V. G. K. Reddi, "Avalanche Breakdown in Transistors," *IEEE Trans. Electron Devices*, **ED-17**, No. 4, pp. 320–335 (1970).

18. B. A. Beatty et al., "Second Breakdown in Power Transistors Due to Avalanche Injection," *IEEE Trans. Electron Devices*, **ED-23**, No. 8, pp. 858–863 (1976).

19. R. H. Winkler, "Thermal Properties of High Power Transistors," *IEEE Trans. Electron Devices*, **ED-14**, No. 5, pp. 260–263 (1967).

20. D. Stolnitz, "Experimental Demonstration and Theory of a Corrective to Second Breakdown in Si Power Transistors," *IEEE Trans. Electron Devices*, **ED-13**, No. 8/9, pp. 643–648 (1966).

21. R. Arnold and D. Zoroglu, "A Quantitative Study of Emitter Ballasting," *IEEE Trans. Electron Devices*, **ED-21**, No. 7, pp. 385–391 (1974).

22. W. Steffe and J. LeGall, "Thermal Switchback in High f_t Epitaxial Transistors," *IEEE Trans. Electron Devices*, **ED-13**, No. 8/9, pp. 635–638 (1966).

23. D. R. Carley et al., "The Overlay Transistor, Part 1, New Geometry Boosts Power," *Electronics*, pp. 71–77, Aug. 23, 1975.

24. K. Demizu and Y. Yamamoto, "Second Breakdown in IC Structured Power Transistors," *IEEE Trans. Electron Devices*, **ED-22**, No. 6, pp. 352–353 (1975).

25. H. C. Chen et al., "Doping Dependence of Second Breakdown in a *p-n* Junction," *Solid State Electron.*, **14**, No. 8, pp. 747–751 (1971).

26. G. Roman, "A Model for Computation of Second Breakdown in Transistors," *Solid State Electron.*, **13**, No. 7, pp. 961–980 (1970).

27. P. L. Hower and P. K. Govil, "Comparison of One- and Two-Dimensional Models of Transistor Thermal Instability," *IEEE Trans. Electron Devices*, **ED-21**, No. 10, pp. 617–623 (1974).

28. R. J. Nienhuis, "Second Breakdown in the Forward and Reverse Base Current Regions," *IEEE Trans. Electron Devices*, **ED-13**, No. 8/9, pp. 655–662 (1966).

29. C. Popescu, "Self-Heating and Thermal Runaway Phenomena in Semiconductor Devices," *Solid State Electron.*, **13**, No. 4, pp. 441–450 (1970).

4.7 PROBLEMS

1. Sketch the output characteristics of an n^+-p-ν-n^+ transistor based on the following assumptions:

   a. The current gain of the transistor is 50 in its active region.

   b. $N_\nu = 10^{14}/\text{cm}^3$ and $W_\nu = 150$ μm. The device cross section is 1 mm $\times$ 1 mm.

   c. The minority carrier lifetime in the ν-region is 10 μsec.

   d. Half the base current is required to provide holes in the ν-region.

2. The transistor of Problem 1 has a collector voltage of 20 V. Compute the width of the depletion layer in the ν-region as the collector current density is slowly increased from zero.

3. Repeat Problem 2 for a collector voltage of 200 V.

4. An n-p-n transistor has a collector-base breakdown voltage of 800 V and a common emitter current gain of 40 at $V_{CB} = 5$ V. Determine the collector voltage at which $\beta_0 = 80$.

5. An n^+-p-n^+ transistor has a base width of 150 μm and a base doping concentration of $10^{14}/\text{cm}^3$. The transistor is biased to $V_{CB} = 500$ V, and it has a common emitter current gain of 40 at low current levels. Compute the value of β_0 for current densities of 50, 100, and 150 A/cm^2.

6. An approximate form of (4.41) is that $x_0 \cong W_B (2 \text{ to } 4\beta_0)^{1/2}$. Show how this expression is derived, and assess the approximations involved.

7. Sketch the critical current densities J_0' and J_0'' for an n^+-p-ν-n^+ transistor as a function of V_{CB}. Thus determine the voltage that marks the boundary between the high field case and the low field case. What is the electric field at this transition point?

The Thyristor

CONTENTS

THE NAME "THYRISTOR" applies to a general family of four-layer devices that exhibit intrinsic regenerative action in their operation. These devices function as switches and can handle large amounts of power by virtue of their ability to hold off high voltages at low reverse currents when in their blocking state, as well as to conduct high currents with low voltage drops in their forward conduction state. The basic thyristor device is the four-layer p-n-p-n structure [1, 2]. In recent years a number of derivatives of this structure have been developed to accommodate various circuit applications. Some of these involve the integration of a pilot trigger thyristor or a shunt diode into the basic p-n-p-n structure. Yet others involve changes in the basic configuration to obtain turn-off capabilities or bidirectional device operation.

At low switching speeds the thyristor is a considerably more efficient switch than a transistor. Thus it has essentially dominated the field of industrial power control, where the operating frequency is usually 60 Hz. At the present time, p-n-p-n devices are available that can switch currents up to 3000 A, at 2000 V, and have surge current capabilities as high as 15,000 A. Recently, however, there has been an increase in the development of circuit applications calling for fast switching devices. Here the transistor becomes more attractive and may be the better alternative.

This chapter emphasizes the device physics of the basic *p-n-p-n* thyristor, since it shares many common aspects with its derivatives. Next, derivative structures are considered, focusing on the additional considerations associated with their operation characteristics and their device design.

The basic *p-n-p-n* thyristor (Fig. 5.1) typically consists of deep *p*-type diffusions made simultaneously into either side of a slice of high resistivity *n*-type silicon, with an alloyed or diffused n^+-type region on one end to form the cathode. An aluminum layer is usually alloyed to the other end of the device to form a p^+-anode contact. An ohmic connection is made to the $P2$ region of Fig. 5.1 to form the gate contact. In essence, therefore, the device has a p^+-p-n-p-n^+ structure. Typical dimensions and surface concentrations are shown for a 2000 V structure.

Junctions J_1 and J_3 are back biased during operation in the reverse direction (cathode positive with respect to anode). Almost all the reverse voltage is supported by J_1, since the regions on either side of J_3 are heavily doped. Thus a small reverse current flows, and the device is said to be in its *reverse blocking mode*. Eventually breakdown occurs at a voltage given by V_{BR}. This is illustrated as 0AB in the device characteristic (Fig. 5.2).

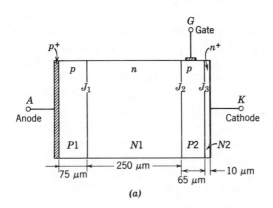

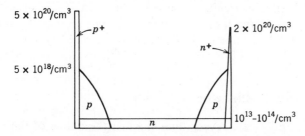

Fig. 5.1   A *p-n-p-n* thyristor.

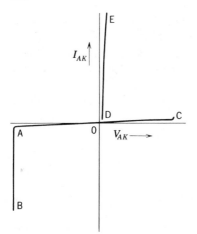

Fig. 5.2   Terminal characteristics for a thyristor.

A somewhat similar situation exists under conditions of forward bias (anode positive with respect to cathode). Now the junction J_2 is reverse biased and supports the forward voltage with very little current flow. The device is said to be in its *forward blocking mode* over this region of its operation (0 C in Fig. 5.2).

Note that during forward blocking conditions, J_1 and J_3 are forward biased, whereas J_2 is reverse biased. Thus at low current levels, $N2P2N1$ behaves very much like a transistor with $N2$ as its emitter. At the same time, $P1N1P2$ also behaves like a transistor, with $P1$ as its emitter.

Injection of base current at $P2$, by way of the gate lead, causes an increase in the forward current between $N1$ and $N2$, by transistor action. This increases the base current of the $P1N1P2$ transistor, which in turn begins to conduct, providing additional base drive for the $N2P2N1$ transistor. This process is thus one of positive feedback. If the gain of the feedback loop is sufficient, the current will grow until the device exhibits a very low voltage drop in its forward direction, even in the absence of any further gate drive at $P2$. The device is now said to be in its *forward conduction mode* (DE in Fig. 5.2).

The action just described persists as long as sufficient load current flows in the anode-cathode circuit. If, however, this current is reduced to a point where the gains of the individual transistors fall so that the regenerative action cannot be sustained, the thyristor will switch back to its nonconducting state. Details of device operation in these various modes follow.

5.1   REVERSE BLOCKING MODE

Junctions J_1 and J_3 are reverse biased when the anode of the thyristor is made negative with respect to its cathode (Fig. 5.1). Since the regions on either side of J_3 are heavily doped, as pointed out earlier, this junction can support only a small reverse voltage (typically 15–25 V) before going into avalanche breakdown. Consequently the device can be modeled as a floating-base $P2N1P1$ transistor (Fig. 5.3), with the cathode connection made directly to the "emitter" region $P2$. Device behavior follows directly from a study of transistor operation in this configuration. Breakdown of this structure may occur by avalanching at J_1. Assuming an abrupt junction, with a heavily doped $P1$-region, the avalanche breakdown voltage is (2.31)

$$BV \cong 5.34 \times 10^{13} N_{N1}^{-3/4} \text{ Volts,} \qquad (5.1)$$

where $N_{N1}(\text{cm}^{-3})$ is the doping concentration of the $N1$-region. This equation must be modified to more closely represent the practical aspects of the situation. Thus for high voltage devices, the p-regions are usually made by deep diffusion. To a good approximation, the doping profile of such a junction is given in exponential form as (2.54)

$$N(x) = N_{N1} \left(e^{-x/\lambda} - 1 \right) \qquad (5.2)$$

Here the origin is taken at the junction, and λ is the space constant. The avalanche breakdown voltage of this junction is more closely approximated by (2.48)

$$BV \cong 9.17 \times 10^9 \, \mathcal{C}^{-0.4} \text{ Volts,} \qquad (5.3)$$

where the grade constant $\mathcal{C}$ is given by N_{N1}/λ and has the dimensions of cm^{-4}.

The $P2N1P1$ transistor can also breakdown by punchthrough of the depletion layer to the $P2$-region. This punchthrough voltage is given for an abrupt junction from (2.26) as

$$V_{PT} = \frac{q N_{N1} W_{N1}^2}{2 \varepsilon \varepsilon_0} \qquad (5.4a)$$

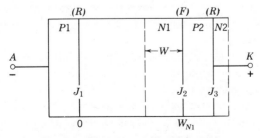

Fig. 5.3 The reverse blocking mode.

where W_{N1} is the width of the $N1$-region. This relationship can also be modified [3] to include the exponential doping profile of the $P1N1$ and $P2N1$ junctions. This can be done by numerical computation, and it results in the inclusion of a correction factor K_1, such that

$$V_{PT} = K_1 \frac{qN_{N1}W_{N1}^2}{2\varepsilon\varepsilon_0} \tag{5.4b}$$

This correction factor (Fig. 5.4) is less than $\pm 10\%$ for practical high voltage junctions.

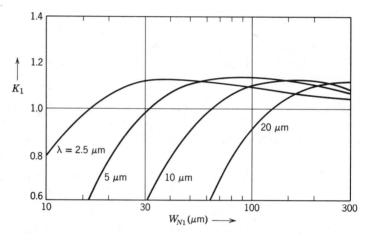

Fig. 5.4 Correction factor for punchthrough voltage. Copyright © 1965 Solid State Electronics. Reprinted with permission from Herlet [3].

By way of example, Fig. 5.5 shows the avalanche breakdown voltage of an exponentially graded p^+-n junction with $\lambda = 10$ μm. Also shown is the punchthrough voltage for the case of $W_{N1} = 200$ μm, so that $K_1 \cong 1.1$. Taken together, these straight lines represent [3] the fundamental limits to the blocking capability of p-n-p-n thyristors having these device parameters. The blocking voltage of the device lies below these limits, depending on the details of its transistor action. Thus Fig. 5.5 must be modified to include these effects.

The actual breakdown of the common emitter p-n-p transistor is achieved when the multiplication factor of J_1 approaches $1/\gamma\alpha_T$, where γ and α_T are the injection efficiency and the base transport factor of the $P2N1P1$ transistor, respectively. The behavior of the multiplication factor for a p^+-n junction has been described by (2.18), so that the reverse

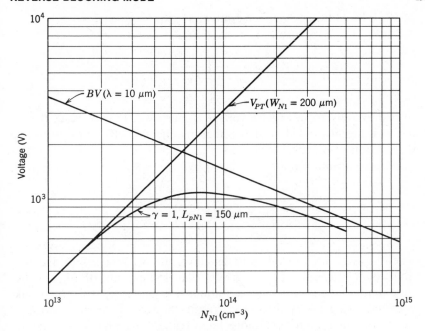

Fig. 5.5   The blocking capability of a thyristor.

breakdown voltage is given by

$$V_{BR} = BV(1 - \gamma\alpha_T)^{1/6} = K_2(BV) \qquad (5.5)$$

where BV is the avalanche breakdown voltage of the $P2N1$ junction. Since the factor K_2 is less than unity, the reverse breakover voltage of the thyristor will be less than the avalanche breakdown voltage of the $P1N1$ junction. A closed form solution of this equation is not possible. However the relative effects of its various terms can be estimated. Thus we can make the following statements:

1. The common base injection efficiency γ is obtained from (4.27) as

$$\gamma = \frac{D_p^*/Q_{N1}}{D_p^*/Q_{N1} + D_n^*/Q_{P2}} \qquad (5.6)$$

where Q_{N1} is taken over W, the undepleted part of the $N1$ region (see Fig. 5.3). In practical situations, W represents a relatively small fraction of W_{N1}. In addition, it is seen from (5.4b) that

$$W = W_{N1}\left[1 - \left(\frac{V}{V_{PT}}\right)^{1/2}\right] \tag{5.7}$$

so that W falls as the reverse voltage is increased. Finally, since the $P2$ region is heavily doped, we know that the injection efficiency is close to unity for most practical situations.

2. The base transport factor α_T is given by

$$\alpha_T = \mathrm{sech}\left(\frac{W}{L_{pN1}}\right) \tag{5.8}$$

where L_{pN1} is the hole diffusion length in the $N1$ region. For a typical high voltage thyristor, L_{pN1} is greater than 150 μm, and W/L_{pN1} is less than unity under reverse bias conditions. Furthermore, W/L_{pN1} shrinks with increasing reverse bias. As a result, the role of the base transport factor becomes more and more important as the reverse voltage approaches the punchthrough limit. Thus the fundamental blocking capability is reduced. Figure 5.6 plots $K_2 = (1 - \gamma\alpha_T)^{1/6}$ as a function of W/L_{pN1}. Thus it represents the ratio of the reverse blocking voltage of the thyristor to the avalanche breakdown voltage of the J_1 junction when $\gamma \cong 1$. The value of K_2 for a non-zero value of injection efficiency ($\gamma = 0.8$) also appears in Fig. 5.6.

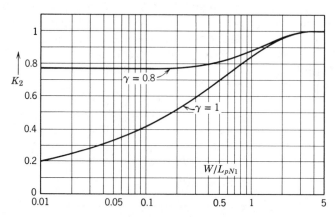

Fig. 5.6 The ratio V_{BR}/BV for a thyristor.

Figure 5.5 shows the actual reverse blocking voltage for a thyristor where $W_{N1} = 200$ µm, $\gamma = 1$, and $L_{pN1} = 150$ µm. This curve was computed by an iteration process, starting with a selection of V_{BR} and using (5.7) as well as Fig. 5.6.

It is interesting to note that the reverse blocking voltage approaches the punchthrough voltage for low values of doping in the $N1$ region. However V_{BR} does not asymptote to BV with increasing doping levels. This is because the ratio of W/L_{pN1} is finite, ensuring that K_2 is always less than unity. In this example, the upper limit of W/L_{pN1} is 1.33 and $K_2 \cong 0.89$. Thus the reverse blocking voltage asymptotes to 89% of the breakdown voltage with increasing doping of the $N1$-region.

Calculation of the blocking voltage for the case of $\gamma \neq 1$ is somewhat complex, since the injection efficiency changes with the width of the undepleted part of $N1$. However a good approximation is that $\gamma \cong 1$ as device operation approaches punchthrough, and $\gamma = \text{constant} < 1$ for operation in the avalanche regime. The situation, rarely encountered in practice, is not discussed further.

It is now possible to estimate the temperature dependence of the reverse breakover voltage, for any fixed value of reverse current. From (5.5), it is seen that the main term of interest is the variation of $(1 - \gamma\alpha_T)$ for the p-n-p transistor. Detailed computations for this term are straightforward, but tedious. We make here an approximate analysis, which provides insight into the physical processes involved. We note that $(1 - \gamma\alpha_T) = \gamma\alpha_T / \beta_{pnp}$. Consequently

$$\frac{d(1 - \gamma\alpha_T)}{dT} = (\gamma\alpha_T)\frac{d}{dT}\left(\frac{1}{\beta_{pnp}}\right) + \frac{1}{\beta_{pnp}}\frac{d}{dt}(\gamma\alpha_T) \tag{5.9}$$

The common emitter gain of the p-n-p transistor becomes large near the breakover point, whereas $\gamma\alpha_T$ approaches unity. Consequently the temperature dependence of $1 - \gamma\alpha_T$ is closely approximated by that of $1/\beta_{pnp}$ in this region.

In transistors such as the $P1N1P2$ device under consideration, the primary temperature dependence is carried by the common emitter injection efficiency. This term is given by the ratio of the injected current density to the space charge generated current density.

It is instructive to study the ratio of the injected current density ($J_{F,\text{diff}}$) to the space charge generated component ($J_{F,\text{rec}}$) as a function of temperature. Figure 5.7 sketches this ratio for a typical high voltage thyristor,*

*Values selected for this figure are a depletion layer width of 0.1 µm for a forward bias of 0.7 V, $D_p = 12.8$ cm^2/sec, $\tau_{sc} = 30$ µsec, $N_{WN1} = 5 \times 10^{13}$/cm^3, and $L_{pN1} = 150$ µm.

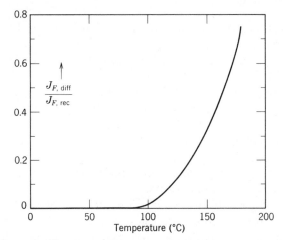

Fig. 5.7   Forward current components versus temperature.

obtained by combining (3.6) and (3.9). The figure indicates that this ratio increases rapidly above 100°C, with a corresponding fall in V_{BR}. Changes below 100°C are negligible, and the dependence of V_{BR} over this temperature range is given by the variation of the avalanche breakdown voltage, which increases at about 0.1% per °C.

   In conclusion, therefore, the reverse breakover voltage increases gradually until about 100°C, then falls off ·rapidly with further increase in temperature. Typically, thyristors are specified around a maximum junction temperature of 125–150°C because of the rapid deterioration of their blocking characteristics beyond this point.

   It is possible to design a thyristor structure with a very wide base, ensuring that the reverse breakover voltage is equal to the avalanche breakdown voltage, and steadily improves with temperature. However this approach results in very poor forward conduction characteristics and is not adopted in practice (see Section 5.3).

5.2   THE FORWARD BLOCKING MODE

Consider a thyristor across which an increasingly forward-bias voltage (anode positive with respect to the cathode) is applied. A fixed gate current I_G is also assumed (Fig. 5.8a). The device is considered to be in its *forward blocking mode*. For this condition, J_1 and J_3 are forward biased (F), and J_2 is reverse biased (R). Thus $P1N1P2$ constitutes a p-n-p transistor with $P1$ as its emitter. Let its current gain be given by α_{pnp}. In like manner,

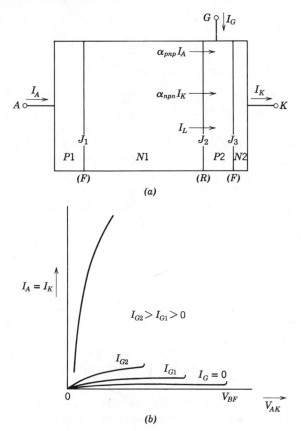

Fig. 5.8 Forward operation of a thyristor.

$N2P2N1$ behaves like an n-p-n transistor ($N2$-emitter) with a current gain of α_{npn}. Figure 5.8a also shows the electron and hole currents for this condition, which are of interest in device operation. To simplify the analysis, we take an injection efficiency of unity for both transistors, and this is an excellent assumption in practical situations.

By transistor action, the anode current I_A results in a hole current $\alpha_{pnp}I_A$ transported across J_2. Furthermore, the cathode current I_K produces an electron current $\alpha_{npn}I_K$ across J_2, in addition to the hole current. Finally, the charge generation current as well as the diffusion-limited leakage current must be included. These are lumped together as I_L, as shown. Then the total anode current is

$$I_A = \alpha_{pnp}I_A + \alpha_{npn}I_K + I_L \tag{5.10}$$

But $I_K = I_G + I_A$, so that

$$I_A = \frac{\alpha_{npn} I_G + I_L}{1 - \alpha_{pnp} - \alpha_{npn}} \tag{5.11}$$

Thus I_A tends to infinity as $\alpha_{pnp} + \alpha_{npn}$ approaches unity. This point marks the termination of the forward blocking mode and the onset of forward conduction.

It is important to note that both α_{pnp} and α_{npn} are strong functions of the current passing through the respective transistors. It is possible to keep the sum of these current gains below unity at extremely low current levels. This is not the case, however, at high current levels.

In actual circuit operation, gate current is absent when the device is in its forward blocking mode. For this situation

$$I_A = \frac{I_L}{1 - \alpha_{pnp} - \alpha_{npn}} \tag{5.12}$$

As long as the leakage current is small, $\alpha_{pnp} + \alpha_{npn}$ can be kept below unity and the device will block. With increasing anode-cathode voltage, however, this current increases, with a corresponding increase in the alphas. Eventually the denominator of (5.12) becomes zero, I_A tends to infinity, and the blocking mode is terminated at a forward blocking voltage V_{BF}.

In practice, the thyristor is held OFF at a voltage below V_{BF} and switched into its ON state by the introduction of a gate current I_G into the base of the n-p-n transistor, thus increasing the electron flow across J_2. This causes an effective widening of the depletion layer on the lightly doped $N1$ side, increasing $(\alpha_T)_{pnp}$ until $\alpha_{pnp} + \alpha_{npn}$ exceeds unity and the device switches to its ON state.*

Figure 5.8b gives a set of blocking characteristics with different values of gate current. Note that the turnover point for these characteristics is more correctly defined [4] as the point at which the $\partial I_A / \partial V_{AK}$ becomes infinite. Thus V_{BF} marks the point at which the sum of the ac or *small signal alphas* reaches unity. This voltage is somewhat lower than that obtained by using the large signal or dc values.

The mechanisms by which the alphas can increase are now considered. During the blocking mode, the forward current is set by the leakage current associated with the reverse-biased junction J_2. Very low values of leakage current cause transistor operation to occur in the ultra-low level

*The details of this process are described in Section 5.4.

injection regime (see Section 3.1.1), where the forward current is dominated largely by recombination in the space charge region of the emitter-base diode. Thus the injection efficiency of both the p-n-p and the n-p-n transistors is initially very low. With increasing applied voltage, however, I_L increases, with a corresponding increase in the ac alphas. Changes in the injection efficiency term are relatively unimportant in high voltage thyristors, except at ultra-low levels of forward current. This is because extreme care is taken to provide for long lifetimes in these devices. This effect must be considered, however, in fast switching devices, which are designed with considerably shorter lifetimes.

A second mechanism is the change in the base transport factor with increasing applied voltage. This effect is primarily significant for the wide base p-n-p transistor. With increasing V_{AK}, the depletion layer across J_2 expands into the $N1$ region, shortening its effective base width. This reflects in an increase in the base transport factor, as described by (5.8). Changes in the base transport factor of the narrow base n-p-n transistor can be ignored here, since this term is essentially unity for all conditions of bias.

Finally, raising the forward voltage across the device results in increasing the multiplication factor as well. This increase occurs for both the p-n-p and the n-p-n transistors, since they share a common collector junction J_2.

The forward breakover voltage V_{BF} is reached when

$$M = \frac{1}{(\gamma\alpha_T)_{pnp} + (\gamma\alpha_T)_{npn}} \qquad (5.13)$$

Comparison with the criterion for reverse breakover $M = 1/(\gamma\alpha_T)_{pnp}$ shows that V_{BF} is always less than V_{BR}. The variation of the forward breakover voltage with temperature thus can be expected to be qualitatively similar to that for V_{BR}, since it is strongly dependent on the temperature behavior of injection efficiency. In this case, however, because both injection efficiencies are involved, the effects of current gain dominate the entire temperature range. Thus the slight increase in breakover voltage, noted in the reverse direction for temperatures below 100°C, is seldom observed here. A typical characteristic, showing the variation of V_{BF} with temperature, is illustrated in Fig. 5.9. Also shown is the variation of V_{BR}, as well as the breakdown voltage for the J_2 junction, for the sake of comparison.

The forward and reverse breakover voltages of a thyristor can be made to approach each other if either (or both) the alphas of the associated transistors can be reduced over the blocking range. One approach is to reduce α_{pnp} by doping the device to obtain a short lifetime in its $N1$ base

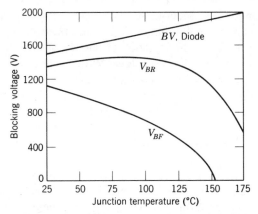

region. Unfortunately, this results in deteriorating the forward conduction characteristics, as Section 5.3 demonstrates. In addition, α_{pnp} now becomes a strong function of the actual value of its base lifetime and is extremely difficult to control during manufacture.

An alternate approach is to provide a technique by which the alpha of one of the transistors (e.g., the n-p-n) is kept extremely small over the entire blocking range. In the ideal situation, α_{npn} should abruptly reach its full value at some critical forward current. Thus the turnover voltage would be set by the rapid excursion of $\alpha_{pnp} + \alpha_{npn}$ through unity at this forward current. One approach to this problem is to use a resistive shunt.

5.2.1 Resistive Shunts

The principle of the resistive shunt is relatively straightforward and can be understood by reference to Fig. 5.10a, where a resistor of value R is connected in shunt with the emitter-base junction of an n-p-n transistor, resulting in a "composite" transistor. For the moment, let us assume that the common base current gain of this transistor is α_{npn} in the absence of the shunt. Consider a forward voltage V_{BE1} across the emitter-base diode. The current gain for this voltage is given (Fig. 5.10b) by

$$\left(\alpha_{npn}\right)_{\text{composite}} = \frac{\alpha_{npn}}{1 + I_2/I_1} \tag{5.14}$$

which can be made much less than α_{npn}. The alpha of the composite

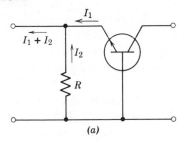

(a)

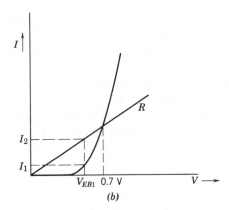

(b)

Fig. 5.10   Resistive shunt.

transistor thus can be made well below unity. However it now rises rapidly with increasing V_{BE}. For example, if a resistor is chosen so that alpha is $0.5\alpha_{npn}$ at $V_{BE}=0.7$ V, the composite alpha will range from $0.047\alpha_{npn}$ at $V_{BE}=0.6$ V to $0.977\alpha_{npn}$ at $V_{BE}=0.8$V.

The resistive shunt is often externally applied in thyristor circuits. A device-oriented approach for providing this shunt, however, is to place a resistive film across the emitter-base junction. The actual implementation of this approach is extremely difficult, because of the need to precisely control the low values of power resistors that must be used here. A more practical approach lies in the use of a cathode short to perform the same function.

5.2.2   Cathode Shorts

The cathode short utilizes the internal lateral resistance of the $P2$ region to provide the resistive shunt path [5]. Thus it is a simple, elegant approach to the problem of controlling the alpha of the n-p-n transistor. The short is provided by making an $N2$ cathode region contact, of width W_K and depth Z, which overlaps the $P2$ gate region by an amount W_G. These regions are the emitter and base of the n-p-n transistor, respectively (Fig. 5.11). Device operation in the forward blocking mode proceeds as follows.

1. At low forward currents, J_1 is forward biased and J_2 reverse biased. All the current from anode to cathode flows through the shorted region, rendering $N2$ practically inoperative as an emitter. The current flow across the device is distorted by the presence of the overlap and has a large lateral component in the $P2$ region. The resulting ohmic drop along this region tends to forward bias J_3, with the highest positive bias occurring at the edge of the contact that is furthest from the short.
2. At some critical forward current for which the positive bias exceeds 0.7 V approximately (for silicon devices), J_3 begins to emit significantly, driving up the alpha of the n-p-n transistor until the sum of the ac alphas exceeds unity, and the thyristor begins to conduct.
3. Once conduction starts, injection proceeds over the entire cross section of J_3, since its transverse resistance to current flow is much less than its lateral resistance.

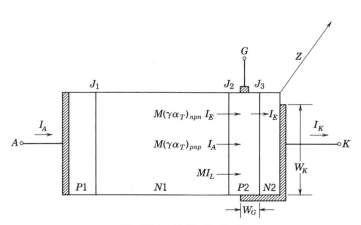

Fig. 5.11   Cathode short.

Figure 5.11 shows the current components of interest just before the start of conduction, and for zero gate current. Here

$$I_A = M\,(\gamma\alpha_T)_{pnp} I_A + M\,(\gamma\alpha_T)_{npn} I_E + M I_L \qquad (5.15)$$

so that

$$I_A = \frac{MI_L}{1 - M(\gamma\alpha_T)_{pnp} - M(\gamma\alpha_T)_{npn}(I_E/I_A)} \tag{5.16}$$

With no cathode short, $I_E = I_A$, so that

$$I_A = \frac{MI_L}{1 - M(\gamma\alpha_T)_{pnp} - M(\gamma\alpha_T)_{npn}} \tag{5.17}$$

Thus V_{BF} occurs when $M = 1/[(\gamma\alpha_T)_{pnp} + (\gamma\alpha_T)_{npn}]$, in accordance with (5.13). This relationship is modified in the presence of the cathode short. From (5.16) it is seen that the anode current is given by

$$I_A = \frac{MI_L}{1 - M(\gamma\alpha_T)_{pnp}} \tag{5.18}$$

if

$$I_E/I_A \ll \frac{1 - M(\gamma\alpha_T)_{pnp}}{M(\gamma\alpha_T)_{npn}} \tag{5.19}$$

For this condition, therefore, the forward breakover voltage becomes equal to V_{BR}. Thus appropriate design of the emitter short allows control of the forward breakover voltage until it becomes equal to the reverse breakover voltage.

Once the thyristor is in forward conduction, it can be turned off by reducing its forward current below the holding value I_H. The magnitude of this holding current can be determined approximately by assuming a critical forward voltage V_{J3} at which the $N2P2$ diode has significant conduction. Then I_H is the current at which the lateral voltage drop in $P2$ becomes less than V_{J3}; that is,

$$I_H \leqslant \frac{V_{J3}Z}{W_K\rho_{P2\square}} \tag{5.20}$$

where $\rho_{P2\square}$ is the sheet resistance of the $P2$ region and $V_{J3} \cong 0.7$ V for silicon devices. The holding current is thus seen to vary inversely as the width of the cathode overlap W_K but is independent of the gate overlap W_G. This result has been borne out by experiment [5].

The turn-on current I_S can also be roughly estimated for a shorted cathode device; it is the value at which the lateral voltage drop in $P2$ exceeds V_{J3}. However there is one important difference between these two processes. The entire $P2$ region is involved during turn-off since J_2 is forward biased. Thus its lateral resistance is relatively independent of the width of the gate overlap. During turn on, however, much of this region is depleted, since J_2 is reverse biased. Because of the diffused nature of $P2$, almost the entire lateral current flow is confined to a narrow region next to J_3, so that the lateral resistance is somewhat higher but decreases as the width of the gate overlap is increased. Thus the lateral resistance is equal to $W_K / Z f(W_G)$, where $f(W_G)$ is a monotonic function of W_G. The turn-on current is then given by

$$I_S \geqslant \frac{V_{J3} f(W_G) Z}{W_K} \tag{5.21}$$

Thus I_S is inversely proportional to the cathode overlap and increases monotonically with the gate overlap. This result has also been confirmed experimentally [5].

The cathode short allows independent control of I_S from values below I_H to values far greater than I_H. Thus it is possible to have a large turn-on current while still maintaining holding currents at the levels encountered in devices without cathode shorts. Figure 5.12 shows the different types of characteristic that can be obtained with such devices.

The cathode short provides a metallic contact to the $P2$ region under conditions of reverse bias. Except for a small decrease in V_{BR}, equal to the breakdown voltage of the heavily doped $P2N2$ junction, it has no effect on the device characteristics in the reverse direction. Under forward bias conditions, however, it results in a V_{BF} characteristic that can be made almost identical to the V_{BR} characteristic. Thus in addition to a higher value of forward breakover voltage, the blocking characteristics exhibits a slight increase with temperature up to about 100°C, followed by rapid falloff beyond this temperature.

In small, diffused devices, the cathode short is usually made by overlapping the cathode metallization over the gate region. In large area devices, however, where the entire slice constitutes a single device, it is common practice to disperse shorts over the whole cathode area, to minimize the concentration of the current density during this turn-on process. The cathode region of a typical high voltage thyristor structure (Fig. 5.13) contains multiple shorts of this type. Both alloying and diffusion technologies (see Chapter 6) can be used for fabricating these structures. The dynamic properties of cathode shorts are discussed in Section 5.5.

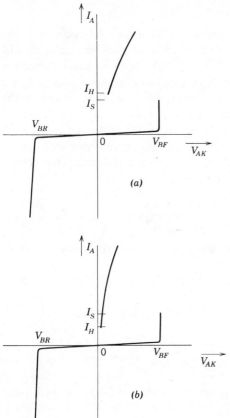

(a)

(b)

Fig. 5.12 Characteristics for thyristors with shorted cathodes.

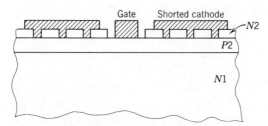

Fig. 5.13 Detail of a shorted cathode structure.

5.2.3 Other Shorted Structures

A thyristor can be built with an anode short (i.e., a short placed across the J_1 junction) in addition to the cathode short. The forward characteristics of this device are quite similar to those of the shorted-emitter structure described previously. In the reverse direction, however, both J_1 and J_3 are shorted so that the maximum reverse voltage that can be sustained is the forward drop across the J_2 junction. Its use is thus restricted to circuit situations that place no reverse voltage requirements on the structure. Such a device is conventionally called a reverse conducting thyristor (RCT).

The forward breakover characteristics of shorted and unshorted thyristor structures can be summarized as follows.

1. In a normal p-n-p-n thyristor, V_{BF} occurs when $M \rightarrow 1/[(\gamma\alpha_T)_{pnp} + (\gamma\alpha_T)_{npn}]$. For such a structure, the forward breakover voltage is well below V_{BR}. In addition, it falls slowly with temperature below 100°C, and rapidly at higher temperatures.
2. In a thyristor with a shorted cathode, V_{BF} is reached when $M \rightarrow 1/(\gamma\alpha_T)_{pnp}$, so that $V_{BF} \cong V_{BR}$. Here the forward breakover voltage rises slightly up to 100°C but deteriorates rapidly at high temperatures.
3. In a reverse conducting thyristor, both J_1 and J_3 are shorted. Consequently the device behaves like a reverse-biased diode ($N1P2$) during the forward blocking mode, until V_{BF} is reached. As a result, the criterion for V_{BF} is that $M \rightarrow \infty$, corresponding to avalanche breakdown of the $N1P2$ junction. Thus V_{BF} for this device is higher than for the other structures. Furthermore, the forward breakover voltage rises gradually at about 0.1% per °C, over the entire operating range of junction temperature [6].

High voltage RCT devices readily operate at junction temperatures of 150°C, as compared to 125°C for the conventional and single-shorted devices. Thus they are capable of handling higher currents in the ON state. An additional benefit that derives from the use of anode shorts is that it is possible to design the device as a reverse biased p-i-n diode when forward blocking. This results in a narrower $N1$ region than its counterpart in a conventional thyristor, thus should display improved conduction characteristics in the forward direction. On the other hand, high voltage devices are somewhat more difficult to fabricate than conventional thyristors. A typical structure, shown in outline in Fig. 5.14, involves the use of a diffused or epitaxially grown n^+-layer to assure that the anode metallization makes a low resistance contact to the $N1$ region. The primary advantage of the RCT structure lies in its dynamic properties (Section 5.5.1.3).

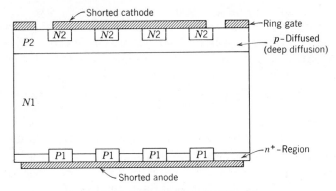

Fig. 5.14 A reverse conducting thyristor.

A thyristor can also be built with a single short placed on the anode side. Such a structure has no advantage over the reverse conducting thyristor, although it shares its nonblocking reverse characteristics. Consequently it is rarely found in practice.

5.3 FORWARD CONDUCTION MODE

The conduction process during the forward blocking mode can be summarized briefly as follows:

1. Holes are injected from $P1$ across the forward-biased junction J_1. They diffuse through $N1$ and are collected across the reverse-biased junction J_2 and into $P2$. Thus they provide excess majority carriers in the base of the $N2P2N1$ transistor.
2. This is accomplished by the injection of electrons from $N2$ across the forward-biased junction J_3. They diffuse through $P2$ and are collected across the reverse-biased junction J_2 and into $N1$. Here they provide excess majority carriers in the base of the $P1N1P2$ transistor.

The situation is thus one of positive feedback, and it is potentially unstable. The transition to instability occurs when the sum of the ac alphas of the associated p-n-p and n-p-n transistors becomes equal to unity, and $I_A \rightarrow \infty$. Alternately, if the current is limited, the voltage across the device will fall and approach zero. Ultimately, however, this fall in voltage drop across the device will limit the flow of minority carriers into the $N1$ and $P2$ regions.

Consider what happens when the entry of holes into $N1$ is limited from the anode side, making it impossible for them to provide for charge neutrality in this region, as shown schematically in Fig. 5.15, where $(1 - \alpha_{pnp})I_A < \alpha_{npn}I_K$. This condition necessitates the back injection of additional holes from $P2$ into $N1$, to preserve charge neutrality.

In like manner, if the supply of electrons injected from $N2$ is insufficient for charge neutrality in the $P2$ region—that is, if $(1 - \alpha_{npn})I_K < \alpha_{pnp}I_A$—electrons will be back injected from $N1$ into $P2$. Either situation results in J_2 becoming forward biased, so that the voltage across its space charge layer collapses.* From this point onward, the junction J_2 plays no further role in device operation, except for its small forward drop, which opposes the forward drops across J_1 and J_3.

The fall in terminal voltage across the thyristor ceases once J_2 becomes forward biased. From here onward, the application of an increasing forward voltage results in increasing the flow of holes from $P1$ and electrons from $N2$, which proceed to flood the $N1$ and $P2$ regions. Thus except for small back-injected components of hole and electron flow across J_2, the device behaves like a p^+-i-n^+ diode (Fig. 5.15b).

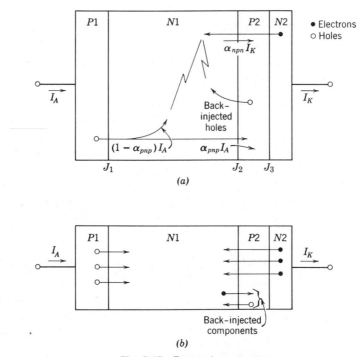

(a)

(b)

Fig. 5.15   Forward conduction.

*A similar process occurs when a transistor goes into saturation. Here, in fact, both transistors $P1N1P2$ and $N2P2N1$ become saturated.

The behavior of the p-n-p-n thyristor in its forward conduction mode is slightly complicated by the presence of an additional, rather heavily doped "base" region ($P2$). The effect of this region on device performance is relatively small, however, because of its extremely narrow width. Since very little recombination occurs in $P2$ during forward conduction, electrons from the $N2$ cathode are essentially injected into the $N1$ region by the transistor action of $N2P2N1$. In effect, then, $N2$ serves as a remote n^+-contact to the $N1$ region.

The characteristics of a thyristor in the forward conduction mode can now be summarized, following the observations made in Section 3.3 for a p^+-i-n^+ diode.

1. The mid-region for the thyristor is $N1$. Conductivity modulation of this region is accomplished by the injection of holes from the anode and electrons from the cathode (by way of the transistor $N2P2N1$).
2. The $N2P2N1$ transistor operates in its saturated mode and has associated with it the voltage drop of such a structure. Conductivity modulation of the $P2$ region is not significant because of its heavy doping.
3. The voltage drop across $N1$ is given by (3.81). Thus the thyristor can behave as a "long" or a "short" p-i-n diode, depending on whether the width of $N1$ is greater or less than the ambipolar diffusion length L_a. In practice, every effort is made to obtain a short diode, by processing for long lifetime in this region.
4. Both carrier-carrier scattering and Auger recombination effects cooperate to reduce L_a at injection levels above $10^{17}/\text{cm}^3$. Thus even if a thyristor is designed to behave like a short structure, it invariably shifts to long at high current densities, especially under surge conditions, where current densities of 1000–3000 A/cm^2 are encountered.
5. Recombination effects in the end regions cause a reduction in the forward current that is available for conductivity modulation of the mid-region, thus increase the mid-region voltage drop. Therefore designers do everything possible to lower the saturation current densities of the $P1N1$ and $P2N2$ diodes.
6. Finally, since both carrier-carrier scatter and Auger recombination set a limit to the high level diffusion length, improvements that can be made by reducing recombination effects in the anode and cathode regions are not as large as predicted by the simple theory.

A number of workers have shown experimentally the similarity between the ON state behavior of the thyristor and the p^+-i-n^+ diode. Some of this work [7] has used optical and potential probing of devices under actual operating conditions. Other work [8], involving the electrical testing of

p-n-p-n and p^+-i-n^+ structures made with identical dimensions and processing, has borne out the foregoing arguments, and today most research on thyristors is undertaken on p^+-i-n^+ diodes, which are easier to fabricate.

5.3.1 The Holding Current

Assume that the forward current through a thyristor is slowly reduced. During this process, the forward voltage drops. At some critical current, however, this voltage suddenly increases as the device switches back to its forward blocking state. This current is known as *the holding current*.

For unshorted thyristors, the holding current is equal to the current at which forward breakover occurs. This is the current at which the sum of the small signal alphas of the associated transistors becomes equal to unity. For any transistor,

$$I_C = \alpha I_E + I_L \qquad (5.22)$$

where I_C and I_E are the dc collector and emitter currents, I_L is the leakage current, and $\alpha (= \gamma \alpha_T)$ is the dc common base current gain. Then the small signal or ac alpha is given by $\tilde{\alpha}$ where

$$\tilde{\alpha} = \gamma \left(\alpha_T + I_E \frac{\partial \alpha_T}{\partial I_E} \right) + \alpha_T I_E \frac{d\gamma}{\partial I_E} \qquad (5.23)$$

Solutions of this equation are relatively straightforward but are not available in closed form. However, numerical computations have been made [9] for a number of transistor structures. In general, the small signal alpha is larger than the dc alpha at low current levels. In addition, both the dc and ac alphas increase more rapidly with current level in transistors which have small ratios of base width to diffusion length.

The holding current of a thyristor is a very sensitive function of the lifetimes of the base regions; numerical computations [10] for this parameter show a reduction of more than two decades as the lifetime is increased by a factor of 2. Consequently it is difficult to control in manufacture. This is not true of the shorted cathode structures, however, where the holding current is a function of the resistivity of the $P2$ base region, as shown by (5.20).

5.4 THE TURN–ON PROCESS

The gate turn-on process of a thyristor can be divided into three separate intervals (Fig. 5.16). The first of these, directly following the application of the gate current, is a turn-on delay phase. An extremely small forward current flows during this interval, so that the sum of the ac alphas is below unity. This phase is terminated at some critical forward current at which $\alpha_{npn} + \alpha_{pnp}$ exceeds unity, followed by an abrupt increase in forward conduction and the start of the rise-time phase. Transistor action plays an important role during this process. As expected, therefore, the critical forward current at which it terminates is related to the length of the gate-cathode emitting edge. A value of 5 A/cm of emitting periphery is typical for modern thyristors.

Let us consider the behavior of a thyristor in its forward blocking mode, to which a gate signal is applied. Initially the $N2P2N1$ transistor begins to conduct. This results in the flow of electrons from $N2$ to $N1$, accompanied by the immediate injection of holes from the $P1$ emitter, as a consequence of charge neutrality. Most of these holes are transported to $P2$ by p-n-p transistor action; therefore both electrons and holes proceed to cross the depletion layer of the blocking junction (J_2).

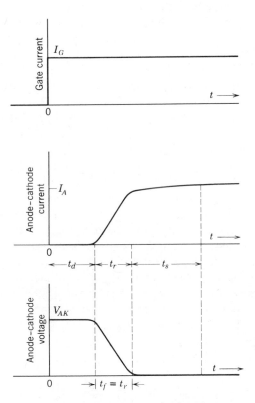

Fig. 5.16   Turn-on of a thyristor.

Since $\alpha_{npn} > \alpha_{pnp}$, it follows that electron flow will predominate across this depletion layer. The effect of this mobile charge in the depletion layer is very similar to that described in Section 4.3.3; that is, it expands on the $N1$ side and shrinks on the $P2$ side. However $\sigma_{P2} \gg \sigma_{N1}$, and the essential effect is therefore one of expansion into the $N1$ region. Thus, one mechanism for termination of the turn-on delay phase of a thyristor is that the injection of gate current brings about an expansion of the space charge layer associated with the blocking junction, so that punchthrough of the $P1N1P2$ transistor eventually occurs. This causes an abrupt increase of forward current, until the sum of the ac alphas exceeds unity. At this point the space charge layer begins to collapse and the rise-time phase is initiated.

In actuality, the critical current that marks the end of the turn-on delay phase is reached somewhat before punchthrough [11, 12]. In part, this is because the initial injection of electrons into the relatively high resistivity $N1$ region sets up a drift field that aids the flow of holes from $P1$. Thus the p-n-p action is considerably faster than it would be if hole transport were by diffusion alone. In addition, the expansion of the depletion layer at J_2 causes a squeezing of the charge in the $N1$ region, toward J_1. The resulting increase in the gradient of the minority carrier concentration at its depletion layer edge corresponds to an increase in the injected anode current. Finally, the gain of the p-n-p transistor increases with the contraction of its base width. All these effects speed the buildup of the forward current to the point at which the rise-time phase begins.

The turn-on delay time t_d is a direct consequence of the processes just described and is always greater than the sum of the base transit times. High voltage thyristors generally have long values of turn-on delay time because of their relatively wide base regions. Moreover, the delay time is dependent on the width of the space charge layer at J_2 before injection of gate current. Hence t_d is reduced for increasing values of V_{AK}, since this causes the space charge region to extend over a greater fraction of the n-base region. In practice, the turn-on delay is generally found to be inversely proportional to the applied gate current; that is, a fixed amount of charge is required for a specific device and in a given bias situation.

The next interval is the rise time t_r, which is initiated once the regenerative process begins. An approximate solution for this term can be obtained by using the principle of charge control. Let Q_{N1} and Q_{P2} be the stored charges in the $N1$ and $P2$ regions, respectively. Then the charge control equations at turn-over are

$$\frac{dQ_{N1}}{dt} = \alpha_{npn} I_K \tag{5.24}$$

and

$$\frac{dQ_{P2}}{dt} = \alpha_{pnp} I_A + I_G \tag{5.25}$$

Recombination effects in both regions can be ignored, since their width is small compared to a diffusion length. Hence

$$\alpha_{npn} I_K = \frac{Q_{P2}}{t_{P2}} \tag{5.26}$$

where t_{P2} is the transit time through the $P2$ region. In like manner,

$$\alpha_{pnp} I_A = \frac{Q_{N1}}{t_{N1}} \tag{5.27}$$

where t_{N1} is the transit time through the $N1$ region. Combining these equations gives

$$\frac{d^2 Q_{N1}}{dt^2} - \frac{Q_{N1}}{t_{N1} t_{P2}} = \frac{I_G}{t_{P2}} \tag{5.28}$$

Solution of this equation results in an exponential growth with a rise time of $(t_{N1} t_{P2})^{1/2}$. The fall time for the anode-to-cathode voltage is also given by this quantity.

The end of the rise-time interval marks the point at which conduction is achieved. This, however, considers the p-n-p-n device as a one-dimensional model, whereas the turn-on process is actually two dimensional. Initially [13] only a small area of the cathode region begins to inject, just as the base current of a transistor causes emitter crowding (see Section 4.3.2). Consequently turn-on is first accomplished over a small area close to the gate contact.

Once this initial area begins to conduct, the external gate loses control. However this highly conducting (but small) region now supplies the necessary forward current to turn on adjacent regions until the conduction process has spread over the entire cross section of the device. Thus although the initial, highly localized turn-on process begins when the sum of the *ac alphas* is unity, the entire process of turn-on is completed only when the sum of the *dc alphas* attains this value. The time interval over which this occurs is referred to as the spreading time t_s.

The end result of this turn-on process depends on the type of circuit in which the thyristor is operating. If the device is fed from a resistive source, current is instantaneously available for passage through the localized turn-on region. This can lead to failure by excessive local heating, as shown in the sequence of Fig. 5.17. Note that the temperature rise of the localized region peaks after the spreading process is initiated. Thus the actual temperature rise is a function of this parameter as well. Intuitively, we realize that a high spreading velocity causes rapid expansion of the localized area, thus reduces the peak temperature attained in the device.

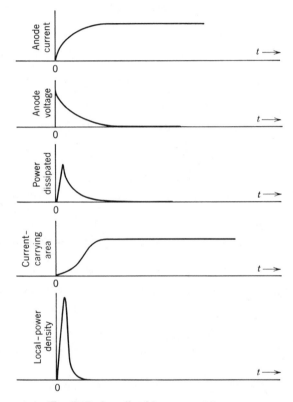

Fig. 5.17   Localized turn-on processes.

5.4.1   The di/dt Limitation

We can now make a crude estimate of the localized temperature rise in a thyristor, in order to determine the device and circuit parameters that are important during this process. We assume that both the anode current and the anode-to-cathode voltage are linear functions of time. The spreading of the conduction process is characterized by a spreading velocity u_s, which is assumed constant throughout the process. A simple concentric structure is considered, with a centrally located gate of radius r_0. During the turn-on interval,

$$i_A = \frac{dI_A}{dt} t \tag{5.29}$$

$$v_{AK} = V_{AK} \left(1 - \frac{t}{t_f} \right) \tag{5.30}$$

and

$$r = r_0 + u_s t \tag{5.31}$$

where V_{AK} is the steady state anode-to-cathode voltage, t_f is the fall time, and i_A, v_{AK}, r are the time-varying values of current, voltage, and spreading radius of the ON state, respectively.

The instantaneous power dissipated is given by

$$i_A v_{AK} = V_{AK} \frac{dI_A}{dt} \left(1 - \frac{t}{t_f} \right) t \tag{5.32}$$

The turned-on area is $\pi[(r_0 + u_s t)^2 - r_0^2]$, so that the power density is given by

$$P = \frac{V_{AK} (dI_A/dt)(1 - t/t_f) t}{\pi \left[(r_0 + u_s t)^2 - r_0^2 \right]} \tag{5.33}$$

Finally, the temperature rise of the hot spot is given by

$$\Delta T = \frac{1}{\rho C} \int_0^\infty P \, dt \tag{5.34}$$

where ρ and C are the density and the specific heat of silicon, respectively.

Inspection of (5.33) shows that this temperature rise is proportional to the anode-to-cathode voltage before turn-on, as well as to the rate of change of the anode current with time. The allowable dI_A/dt for a specified V_{AK} is thus an important rating for any thyristor and must not be exceeded in circuit applications. It is usually specified for a variety of repetitive switching conditions commonly encountered in practice.

5.4.1.1 The Spreading Velocity

Since we know from (5.33) that the temperature rise is also related in an inverse manner to the spreading velocity, it is of interest to understand the behavior of this parameter. Theories for the spread of conduction in a thyristor are at a very rudimentary stage. One approach [14] to this problem has been based on the assumption that the initial conduction is a high level process, which means that the $P2$ region consists of equal electron and hole densities and is essentially neutral. This allows the propagation of the "plasma" to be treated as a purely diffusional process. Following this approach, it has been calculated that the spreading velocity varies as $J^{1/n}$, where n ranges from 2 to 6.

An alternate approach [15] is to consider that the initial conduction is not a high level process, as perhaps is the case if the $N2$ emitter has a high injection efficiency. In this situation a lateral electric field is created in the base region by the transport of majority carriers from the gate to the cathode edge. Thus the spread of conduction can be treated as a drift process because of this lateral field.

The drift model is attractive because of its simplicity. We compute the effective emitter edge of the $N2$ region, as was done for a transistor in Section 4.3.2. Manipulation of (4.41) results in an effective emitting edge of x_0, whose magnitude is given to a very rough approximation by

$$x_0 \cong W_{P2} \left(2 \text{ to } 4\beta_0\right)^{1/2} \tag{5.35}$$

where W_{P2} is the effective width of the $P2$ region and β_0 is the common emitter current gain of the $N2P2N1$ transistor.

The voltage drop across this emitting edge can be chosen as $3kT/q$ in order for emitter conduction to be down to 5% of its peak value. The lateral electric field is thus given, to a crude approximation, by $3kT/2qx_0$. For typical thyristor dimensions, this is about 3–8 V/cm, and results in a spreading velocity of 3000–8000 cm/sec (assuming an average mobility of 1000 cm^2/V-sec).

The drift model has generally been favored over the more complex diffusion model. However, it is probably the less physically correct of the two. Certainly, the high level diffusion model is more appropriate for modern designs [16], which utilize high gate drives to obtain a more uniform turn-on.

A number of experimental techniques have been used to measure the spreading velocity. These include the use of probing techniques [17], infrared imaging [18], and microwave reflection [19]. The results obtained by all workers have been relatively consistent. These are now summarized for concentric geometry structures with a central gate contact [16] and high gate drive currents:

1. $u_s \propto (J)^{1/n}$, where J is the current density and $n = 2 - 3$. Typical values of u_s are 3000 cm/sec at a current density of 20 A/cm^2, increasing to 10,000 cm/sec at 700 A/cm^2.
2. $u_s \propto \tau^{1/2}$, where τ is the average lifetime of the base regions. The spreading velocity typically ranges from 3000 cm/sec for short lifetime (1μsec) thyristors to 20,000 cm/sec for long lifetime (20 μsec) structures.
3. $u_s \propto 1/W_{N1}$, measured over devices with base widths from 100 to 800 μm (u_s values from 8000 to 1000 cm/sec, respectively). Combination of this equation with the last shows that the propagation velocity is inversely proportional to the W/L ratio of the p-n-p transistor.

5.4.1.2 Improvement of the di/dt Rating

Increase of the di/dt capability can be readily accomplished by overdriving the gate. This appears to enlarge the initial turned-on region, causing a local reduction of the current density at this crucial point in the process. Furthermore, a high gate drive eliminates the effects that minor material or process inhomogeneities might have on the spreading velocity.

Device-oriented techniques for improving the di/dt rating of thyristors have been based either on increasing the spreading velocity or on increasing the initial turn-on area. One approach for increasing the spreading velocity is to decrease the W/L ratio for the $N1$ region, as suggested earlier. This can be done by using a longer lifetime process or by reducing the length of the $N1$ region. However both these approaches must be used with care because they affect other operating characteristics.

Another approach is illustrated in Fig. 5.18. Here [20], by using a small contact on the cathode, it is possible to develop a lateral voltage drop from a to b, during the early phases of conduction. This sets up a lateral electric field in the $N2$ region, which aids in spreading of the conduction process to the entire cathode.

Yet another approach is to design the device so that no part of the cathode is greater than a certain maximum allowed distance from the gate electrode. This mandates the use of a gate electrode with a large perimeter, ensuring that the adjacent cathode area is as large as possible.

A number of interdigitated designs have been developed to achieve this goal. Perhaps the most effective is a structure in which both the gate and the cathode have an involute pattern [21]. Figure 5.19a presents the geometric aspects of one arm (gate or cathode) of such a structure.

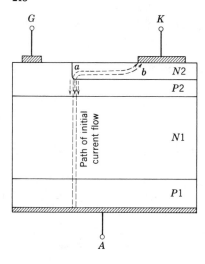

Fig. 5.18 Speed-up technique for plasma propagation.

Consider a string that is wound around a circle of radius r_0. Any point on this string will generate an involute when it is unwound tightly. The generating circle is known as an evolute, and any number of involutes can be formed by suitable choice of starting points on the perimeter of the generating evolute. The path ABC in Fig. 5.19a is that of one involute, whereas DEF is that of a second. The polar equation for ABC is

$$r = r_0(1 + \theta^2)^{1/2} \tag{5.36}$$

and that for DEF is

$$r = r_0 \left[1 + (\theta - \delta)^2 \right]^{1/2} \tag{5.37}$$

where δ is as shown.

The unique property of any two involutes generated from the same evolute is that they are always equidistant from each other, by a fixed amount equal to $r_0\delta$. It follows then that by choosing suitable values of δ, it is possible to make any number of gate and cathode fingers, each of uniform thickness, while preserving a constant separation between them throughout the wafer. Extremely uniform turn-on can be effected in this manner, with a correspondingly increased di/dt rating. A detail of this structure appears in Fig. 5.19b.

Involute gate structures have been made with as many as 36 gate and cathode fingers, resulting in an improvement in the di/dt capability of thyristors by a factor of as much as 25. Modern devices of this type can be designed to have di/dt ratings in excess of 1000 A/μsec.

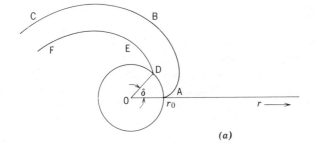

(a)

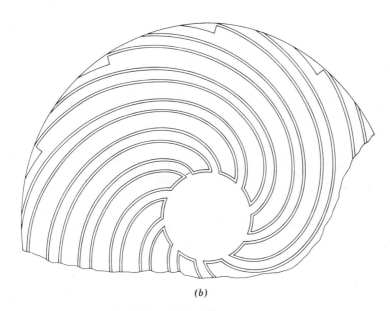

(b)

Fig. 5.19   The involute structure.

Involute and other area control techniques share the disadvantage of necessitating the use of high resolution photolithography which leads to increased cost in manufacture. In addition, the effective area of the device is reduced, as is its current-handling capability. Finally, the gate current required for turn-on is increased by as much as a factor of 10, because of the comparable increase in the length of the emitting periphery. Often this necessitates the use of an auxiliary thyristor to provide gate current for the main device. The amplifying gate thyristor provides this function in an integrated fashion.

5.4.1.3 The Amplifying Gate Thyristor

Figure 5.20*a* is a cross-sectional view of an amplifying gate thyristor* [22]; the main thyristor is indicated by $P1N1P2N2$. An additional annular cathode diffusion makes up the necessary auxiliary thyristor $P1N1P2N2'$, with a shorted emitter connection provided by the cathode contact overlap K'. In operation, this auxiliary thyristor turns on first, by means of a small gate current. This early turn-on is inherent, since its cathode is closest to the gate contact. The cathode current of this device serves as the gate current for the main thyristor, because of the cathode short K'. Thus, the $P1N1P2N2'$ device acts to amplify the gate current and supplies a large

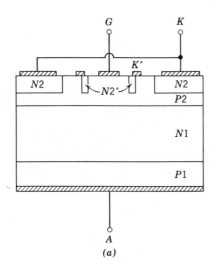

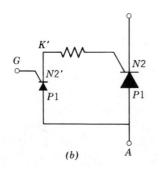

(b)   A   **Fig. 5.20   The amplifying gate thyristor.**

*Sometimes referred to as a regenerative gate thyristor. Patent No: 3,486,088 by D. I. Gray et al.

current to turn on the main thyristor. The equivalent circuit of this combination is given in Fig. 5.20b.

Amplifying gate structures thus provide two important features: first, a large overdrive current, which serves to increase the initial turned-on area; second, the high drive currents that are necessary for the operation of structures with large gate perimeters.

A common problem encountered with amplifying gate thyristors is premature turn-on of the main thyristor before the auxiliary one has fired. This is sometimes caused by local inhomogeneities in the silicon material or by crystallographic imperfections introduced during crystal growth or device processing. One means of avoiding premature firing is by control of the lateral voltage drops under N_2 and N_2'. It has also been proposed [23] to adjust the crystal growth conditions so that its central core is more heavily doped than the surrounding region. In this way, the forward breakover voltage of the inner device can be made intentionally lower than that of the main thyristor, ensuring that the inner device always fires first.

5.4.2   The dv/dt Effect

It has been observed that under transient conditions, a blocking thyristor can switch to its forward conduction state at voltages well below the breakover voltage. The degree of this degradation is dependent both on the magnitude of the anode voltage and its rate of increase. This phenomenon is termed the rate of rise effect, or the dv/dt effect.

During forward blocking, the junction J_2 of the thyristor is reverse biased and has a depletion layer associated with it. In steady state, the forward current was shown to be caused by charge generation in this depletion layer, by the diffusion-limited leakage current, and by the effect of transistor action. Under transient conditions it is possible to pass in addition a displacement current across this junction, of magnitude $C(dv/dt)$, where C is the junction capacitance. Thus if the sum of the ac alphas exceeds unity during this transient phase, premature switching into forward conduction can come about.

5.4.2.1   Improvement of the dv/dt Rating

One approach to improving the dv/dt capability of a thyristor is to reverse bias the gate-cathode circuit.* This ensures that the displacement current is drawn from the gate and will not affect the gains of the individual transistors. A second method is to reduce the lifetime of the $N1$ and $P2$

---

*This approach is treated more fully in discussing the turn-off process.

regions, although device performance in the forward conduction mode is degraded thereby.

A very effective technique is to use a shorted cathode structure. In this way, the capacitive displacement current is shunted around the emitter junction and does not affect the current gain of the n-p-n transistor, except at extremely high levels. The use of multiple shorts, as in Fig. 5.13, allows large lateral currents to flow in $P2$ without forward biasing the $N2P2$ junction. Thus the dv/dt capability of the device increases as the number of such shorts is increased. Typically, anywhere from 6 to 100 shorts may be incorporated into a device, depending on the design requirements.

It can be shown that the voltage drop between two shorts is related to the value of dv/dt by [24]

$$V_0 = C_0 \rho_{P2\square} \frac{dv}{dt} R \qquad (5.38)$$

where

$$R = \frac{1}{16}\left[d^2 + D^2 \left(2\ln\frac{D}{d} - 1 \right) \right] \qquad (5.39)$$

Here, C_0 = capacitance per unit area of the J_2 junction at the desired blocking voltage
$\rho_{P2\square}$ = sheet resistance of the undepleted $P2$ region
d = diameter of the shorts
D = distance between shorts (center to center)

For the short to be effective, V_0 must be kept to a maximum of about 0.5 to 0.6 V, to avoid forward bias of the $N2P2$ junction.

A study of (5.38) and (5.39) shows that the dv/dt capability of the thyristor goes up with the number of shorts that are used, rising quite rapidly for a large number of shorts (as many as 100 shorts are found in some designs). On the other hand, these shorts lead to a reduction in the effective device area, hence in its current-carrying capability.

It is worth remembering that the static behavior of the thyristor is also improved by the presence of emitter shorts. Here, however, almost the full improvement in V_{BF} can be achieved by the insertion of a few shorts (e.g., six).

The use of multiple shorts can result in an improvement in the dv/dt capability of thyristors by a factor of 100 or more. Typically, 20 V/μsec is achieved in devices without shorts. On the other hand, shorted devices with dv/dt capability from 200 to 2000 V/μsec are now available.

5.4.2.2 Interaction with the Turn-on Process

The use of cathode shorts has a number of attendant disadvantages. For example, they take up space, producing a corresponding reduction in the active area of the device. As a result, the forward voltage drop for any specified current density is increased. In addition, cathode shorts act to shunt lateral current out of the $P2$ base, reducing the lateral electric field, and hence the spreading velocity. Finally, they tend to reduce the $N2P2$ junction voltage by shunting some of the load current around it.

From (5.38) it is seen that the voltage drop between shorts varies linearly with R, where R is a parameter characterizing the shorting pattern. Experiments with thyristors having high gate drive have shown [16] that the spreading velocity does indeed increase as R increases, but with a logarithmic dependence (i.e., $R \propto \ln u_s$). Typical values (for a current density of 100 A/cm^2) range from 3000 cm/sec for $R = 1 \times 10^{-3}$cm^2, to 10,000 cm/sec for $R = 9 \times 10^{-3}$cm^2.

A variety of geometrical patterns have been proposed for emitter shorts. Of these, the triangle equilateral is most commonly used and results in equal spacings between each short and its nearest neighbor [25]. In addition, shorts that are close to the gate participate in the propagation process to a greater extent than those further away. Consequently their role is more crucial to the turn-on process when gate triggering is used. Typically, the initial conducting area tends to occur in the vicinity of the center of an elementary cell of the short array and is particularly restricted in area. It is therefore desirable for the turn-on process to be initiated in a control region, before spreading over the entire shorted-emitter region.

In summary, we note that there is significant interaction between the shorts and the turn-on mechanism of a thyristor. In particular, the use of cathode shorts improves the dv/dt and the high temperature blocking capabilities of the device, but it also affects adversely the injection properties of regions located between the shorts, and hence the di/dt rating. The use of shorts cannot be indiscriminate, and care must be exercised in placing them at locations where they are fully effective. Suitable placement, coupled with high gate drive currents, has resulted [26] in the development of devices with di/dt ratings in excess of 1000 A/μsec, while still maintaining a high dv/dt capability (> 1000 V/μsec).

5.4.3 Optical Turn-on

Gate currents for a thyristor can also be provided by the optical generation of hole-electron pairs in the device. This is done by irradiating the emitter and underlying adjacent base region. Devices that can be triggered in this manner are usually of low power handling capacity and are suitable for use in applications where electrical isolation is important. Often they are used to provide the gate current for larger thyristors, that is, in an externally connected amplifying gate configuration, with the light-activated device serving as a pilot thyristor.

The turn-on of a thyristor by optical means is an especially attractive approach for devices that are to be used in extremely high voltage circuits. A typical applications area is in switches for dc transmission lines operating in the hundreds of kilovolt range, which use series connections of many devices, each of which must be triggered on command. Optical firing in this application is ideal for providing the electrical isolation for trigger circuits required with these devices. Conventionally, this has been done with the aid of a floating photodetector-amplifier configuration, or with a small optically controlled thyristor as a pilot device, electrically connected to the thyristor's terminals. However the optimum solution lies in the design of devices that can be fired directly from an optical source, since this avoids the necessity of using (and interconnecting) auxiliary devices in extremely high voltage circuits.

A complete thyristor system of this type includes light sources, fiber optic cables, and optically activated thyristors. Considerable information on light sources and fiber optic light guides is available in the literature [27, 28]. This section therefore concentrates on the device aspects of thyristors for use with these components.

At the present time, the gallium arsenide light-emitting diode is the most practical light source. The gallium-arsenide/gallium-aluminum-arsenide double heterojunction laser [29], a promising candidate for the near future, provides a considerably greater power output. Both these sources emit light in the 0.8 to 0.95 μm range and have a penetration depth of 10–50 μm for silicon. Thus carrier generation in thyristors is not limited to surface layers but can penetrate into the base of the $N2P2N1$ transistor as well.

It should be noted that the actual effect of photogeneration in the $P2$ base is twofold. Electrons generated in this region are primarily transported across the $N1P2$ depletion layer because of the strong aiding electric field. This electron current provides base drive for the p-n-p transistor. Holes generated in the $P2$ region, however, provide direct base drive from the n-p-n transistor. This term dominates the turn-on process, since $\alpha_{pnp} < \alpha_{npn}$. Optical firing is thus very similar to gate triggering, in that both techniques provide for increasing the gating current to the n-p-n part of the thyristor.

Typically, optical power is directly delivered to a small central portion of the $N2$ emitter by means of a light pipe. The rest of the $N2$ region consists of the main thyristor cathode. Sometimes auxiliary amplifying gate devices are also incorporated in this region.

A number of special problems arise in the design of optically triggered thyristors, primarily because the power output of optical sources is extremely low. In addition, coupling in and out of the optical fiber is quite inefficient, especially in multimode systems with light-emitting diodes. Finally, transmission losses through the fiber tend to be quite high.

For modern light-emitting diode systems, the incident light power on the thyristor is in the 5 mW range and results in a few milliamperes of hole current in the $P2$ base region. A continuous wave laser, on the other hand, is capable of providing about 5 times as much incident power on the thyristor. The development of improved optical signal delivery systems is thus an important direction for further work in this area.

The main requirement for an optically triggered thyristor is high sensitivity. Moreover, it must have high di/dt and dv/dt capability if it is to operate reliably. High sensitivity and high di/dt capability can be met simultaneously by the use of integral amplifying gates. In fact, designs have even been developed for devices with two stages of internal gate amplification [30].

The high dv/dt requirement can be met by using shorted emitter structures. Unfortunately, these shunt both the capacitively generated hole current and the optically generated component. Thus their use directly contradicts the demands for low trigger levels, and it is necessary to design the shorting structure to effect an optimum tradeoff between these requirements. However, dv/dt triggering can also be prevented if alternate means are adopted for reducing the forward bias across the cathode-gate junction at any specific value of gate current. One such approach is to electrically raise the potential of the optically fired cathode at the same time that the p-base potential is raised. A technique for accomplishing this is to connect its $N2$ region to the $P2$ base by means of a high resistance, so that it follows the base potential (Fig. 5.21a). Now, any dv/dt capacitive current, or forward blocking current, is diverted around the optically fired thyristor, keeping its sensitivity high. The rest of the structure is heavily shorted and may also incorporate amplifying thyristors, if required.

A practical realization of this compensation scheme is presented in Fig. 5.21b. Here a ring contact is placed around the device and connected to the optically fired auxiliary thyristor. A high resistance is provided in this path by etching of a groove in the $P2$ base. In this manner the $P2$ base is electrically connected to the cathode of the optically fired thyristor through this resistive path.

Structures of this type have been developed with dv/dt ratings of as high as 400 V/μsec when triggered with 5 mW of incident power. The dv/dt rating for such devices falls to 50 V/μsec when the compensation potential is disconnected from the auxiliary thyristor, indicating the effectiveness of this technique. Thus it has great promise for solving the basic problems

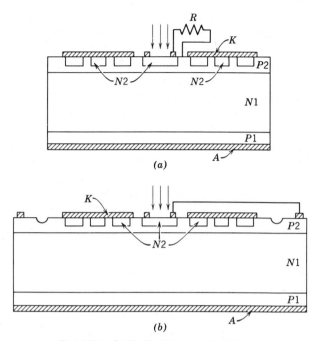

Fig. 5.21 Optically triggered thyristor.

associated with meeting the somewhat conflicting requirements of optically triggered high voltage thyristors.

5.5 REVERSE RECOVERY

A distinction must be made between turn-off and reverse recovery in a thyristor. Turn-off in a thyristor requires that its forward current be reduced to zero. Full reverse recovery, on the other hand, occurs only when the device is brought from its forward conduction state to its forward blocking state, upon reapplication of the supply voltage. Additionally, this requires that it has fully recovered its dv/dt rating; that is, accidental turn-on will not occur because of a rapid transient within the allowable value of this parameter. It is customary to define this "full reverse recovery" time as the turn-off time, since it is the relevant parameter in circuit applications.

The simplest technique for turning off a thyristor is to reduce its forward current to a value below the holding current. Now the stored charge in the bases must decay by recombination alone, to a point at which the sum of the ac alphas becomes less than unity. A second approach is to interrupt the load circuit. Here again, charge decay is purely by recombination. Both these processes are inherently slow, because of the long lifetime that is usually required to obtain good forward characteristics. In either situation, if the carrier density remaining in the device is above some critical value, the reapplication of forward voltage will cause the device to fail by conducting when it should be in its forward blocking mode.

Customarily, a thyristor is turned off by applying a negative voltage to the anode-cathode circuit. This, in turn, develops a reverse current by which some of the stored charge can be removed, thus speeding the overall recovery process. Details of the reverse recovery process are considered for the idealized situation (from a circuit designer's viewpoint), where a reverse voltage V_{REV} is applied to the anode-cathode circuit of a forward conducting thyristor through a resistance of value R, at $t = 0$. Figure 5.22 shows

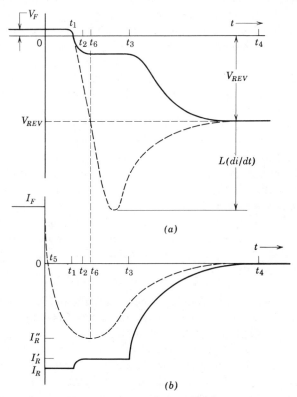

Fig. 5.22 Turn-off wave forms.

current and voltage wave forms for the recovery process, which proceeds as follows:

1. $0-t_1$: During this period the full reverse current $I_R = V_{REV}/R$ is carried by the diode as the carrier concentration in the base falls toward zero (see Fig. 5.22). This process is very similar to what is initially seen in the recovery characteristics of a p^+-i-n^+ diode. Here, however, the carrier concentration becomes zero at the J_3 junction first, since the minority carrier lifetime of the $P2$ region is shorter than that of $N1$. In addition, carrier removal in $P2$ is aided by the gate circuit to which it is connected, as well as by injection out of this region and into $N1$. The voltage drop across the thyristor is positive during this interval and just begins to reverse at $t = t_1$.

2. t_1-t_2: During this interval the depletion layer builds up across the J_3 junction. Again the procedure is analogous to that in a p^+-i-n^+ diode. However this process is terminated by avalanche breakdown of J_3, which typically occurs at 15–25 V. Reverse voltages used to assist in the recovery process are normally well in excess of this value.

3. t_2-t_3: During this interval the reverse current is given by $I_R' = (V_{REV} - BV_{J3})/R$, until the minority carrier concentration at the edge of J_1 falls to zero, so that *this* junction becomes reverse biased. This "flat-top" region is of relatively short duration.

4. t_3-t_4: From this point onward, the device behaves like a floating-base p-n-p transistor with $P2$ as its emitter. During this interval the $P2$ region injects holes into $N1$. At the same time, holes in this region are removed by transport across the depletion layer that builds up at J_1. Furthermore, because of the low electron injection efficiency of the $N1P2$ junction, the removal rate for electrons by this path is relatively low. Consequently the decay of carriers in this region comes about by the sweeping action provided by expansion of the space charge region that develops at J_1, and by minority carrier recombination in the neutral part of the $N1$ region. These processes continue until the reverse current decays to a value low enough to allow the device to block in the forward direction.

The mathematical treatment of the dynamics of the process just described follows closely the analyses for diode behavior in its various regions of operation and has been detailed [32] elsewhere. In practical circuits, however, these effects are dominated by inductive effects in the load circuit [33]. These inductances either are parasitic or are intentionally placed to reduce the di/dt of the load circuit. In either event, they do not allow the immediate reversal of current from I_F to I_R. In addition, a large

$L(di/dt)$ voltage is built up across the device terminals. The actual details of the reverse characteristic are shown by the dashed lines in Fig. 5.22.

Here the first effect of the circuit inductance is that the current cannot be reversed instantaneously. Instead it falls almost linearly with time and becomes zero at $t = t_5$. Thus the flat-top region at I_R is masked by this process.

Next, the reverse current peaks at t_6, at a value less than I_R, (and possibly less than I_R', depending on the device lifetime). At this point the voltage across the circuit inductance becomes zero, which means that the full reverse voltage V_{REV} is present across the thyristor. The reverse voltage across the thyristor continues to increase, reaching a peak value of $L(di/dt)$ in excess of V_{REV}, eventually decaying to zero with the reverse current. This final decay phase is essentially similar to that for the resistive load condition.

In many thyristor applications (e.g., inverters) a common circuit configuration requires that a diode be placed in antiparallel with the thyristor (n-side tied to anode, and p-side to cathode). This severely restricts the reverse overvoltage, in addition to creating an alternate path for reverse current flow. Thus the reverse recovery current has a greatly reduced peak, and a longer duration.

The time t_4 marks the point at which the device is in its reverse blocking mode. Here J_1 is forward biased, J_3 is in avalanche breakdown, and J_2 is reverse biased. At this point a substantial amount of excess charge is stored in the base regions of the thyristor. This charge will fully decay to zero if left for a sufficiently long time. In practical applications, however, forward voltage is reapplied (anode terminal positive with respect to the cathode) before all this charge has decayed. This causes a flow of holes across J_2 and to the cathode, resulting in an exponentially decaying spike of forward recovery current [34], which is somewhat similar to the reverse recovery current.

The magnitude of this forward recovery current increases with the dv/dt of the reapplied anode voltage; that is, it behaves much like a dv/dt induced displacement current. Its collection at the cathode raises the bias on J_3 and the injected current in the $N2P2N1$ transistor. If sufficiently large that the sum of the ac alphas exceeds unity, the device will fail to achieve its forward blocking state. Full recovery is achieved only if the forward recovery current is unable to turn on the thyristor accidentally.

Experimentally, the forward recovery current has been found to increase with junction temperature. The time taken for a thyristor to fully recover at 125°C is typically about twice the recovery time at 25°C, primarily because of the increase of minority carrier lifetime with temperature.

Temperature effects are also important on a localized basis. Thus if the thyristor is accidentally switched on before recovery, this self-switching will occur near the gate where localized temperatures are higher than in the surrounding regions. The actual temperature in this region is a function of such operating parameters as duty cycle and repetition rate.

The magnitude of the forward recovery current is also found to increase with carrier lifetime, with the reapplied dv/dt, and with the rate of commutation of the anode current. It is thus a sensitive function of the circuit conditions under which the device must operate.

5.5.1 Improvement of Recovery Characteristics

We have noted that the carrier density in the base regions must fall below some critical value if the device is to remain in its blocking state when the forward voltage is reapplied. Thus techniques for improving the recovery characteristics of thyristors must depend on the rapid annihilation or removal of excess carriers from these base regions. Both circuit and device-oriented approaches can be used for this purpose.

Let I_0 be the magnitude of the forward recovery current spike, which is assumed to decay exponentially with a characteristic time constant that is related to the minority carrier lifetime. Then its magnitude will fall below the holding current I_H, in a time interval proportional to $\tau \ln(I_0/I_H)$. Consequently the recovery time is proportional to this term. Increase of the holding current thus represents one technique to lengthen the turn-off time. Unfortunately this approach is not available to the designer, since the device is already heavily shorted. Furthermore, dependence on the holding current is logarithmic, thus relatively weak.

5.5.1.1 Lifetime Reduction Techniques

A second means of improving the recovery characteristic is to reduce the minority carrier lifetime in the $N1$ and $P2$ regions. Typically lifetime reduction by a factor of K serves to reduce the recovery time by a factor of about $K/10$.

The use of short lifetimes in a thyristor is accompanied by some harmful side effects on device performance. First, the forward voltage drop across the device depends on $(d/L)^2$ and is increased for any specific current. In addition, the spreading velocity varies with the square root of lifetime, as seen in Section 5.4.1.1. Thus both the steady state and the transient current-handling capability of the thyristor are deteriorated by this process.

A number of processes for lifetime control are detailed in Chapter 6; the unique advantages and disadvantages of each are summarized in this section. Gold doping is the most commonly used technique for lifetime control. In practice, it is done by evaporating a film of gold on the device, followed by diffusion. Gold is a rapid diffuser, and all regions of the device are affected by this process. However the gold doping is not uniform throughout the device and is more concentrated in regions of high damage (i.e., in the highly doped regions). Additionally, gold tends to accumulate preferentially in phosphorus-doped regions because of enhanced solid solubility effects due to ion pairing. As a result, gold diffusion is a very difficult process to control in a reproducible manner. Finally, the use of gold results in an increase in the charge generation component of the leakage current, since its dominant energy level is located close to the center of the band gap (for n-type silicon). Thus the forward breakover voltage of a thyristor is greatly reduced by gold doping.

Recently it has been noted that platinum gives rise to asymmetric energy levels in silicon; therefore this element can be used for lifetime control in the neutral regions without reducing the lifetime in the space charge region. Consequently its use results in low leakage but has little effect on the forward breakover voltage. The technology of platinum doping is relatively primitive. Platinum-doping is performed by diffusion from a compound of the impurity, which is usually placed on the slice by a spin-on doping technique. The detailed mechanism of platinum diffusion in silicon is not well understood; qualitatively, however, its behavior is similar to that of gold.

Electron irradiation in the 1–2 MeV range can also be used to control lifetime without increasing the leakage current. This process has other significant advantages: namely, it is conducted on finished devices, at room temperature, followed by a low temperature anneal cycle. Consequently lifetime can be precisely controlled to any desired value.

5.5.1.2 Gate-Assisted Turn-off

It is common circuit practice to apply a reverse bias between the gate and the cathode during the turn-off phase. This results in a reduction in the turn-off time (for full recovery) by a factor of 2 to 3, representing a significant improvement in device performance. This improvement comes about because the reverse-biased gate acts to divert the forward recovery current that would otherwise flow through to the cathode during the reapplication of the supply voltage. As pointed out earlier, this current can cause the thyristor to fail by turning on, if it exceeds some critical value.

Current flow during this phase of turn-off is shown in Fig. 5.23a for the case of no gate assist, and in Fig. 5.23b for gate-assisted operation. Note that although the use of a negative gate bias diverts the forward recovery current around the cathode junction, lateral current flow in the $P2$ base

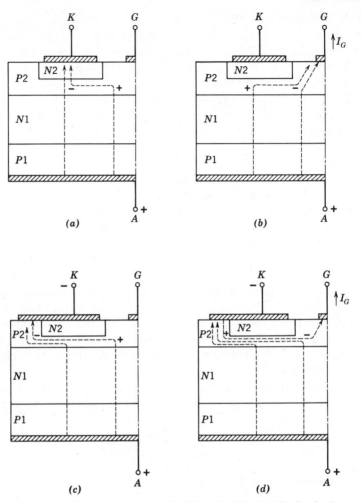

Fig. 5.23   Gate-assisted turn-off with and without cathode shorts.

results in a voltage drop that forward biases J_3. This can *still* lead to forward conduction, if the forward bias is sufficiently large. Consequently devices designed for gate-assisted turn-off should have a low lateral resistance in the $P2$ region, to reduce this voltage drop to a value below that required for conduction across J_3 (i.e., below 0.5–0.6 V).

Ordinarily, operation of a thyristor in the gate-assisted mode precludes the use of heavily shorted structures, since these shorts prevent the application of a significant reverse bias (10 V is typical) across the gate-cathode terminals. Nevertheless such structures (with all their attendant advantages) can be used [35] if the device is specifically designed for gate-assisted turn-off.

Figures 5.23c and 5.23d show the current flow paths with and without gate assist, for the case of a shorted-cathode thyristor. Note that the lateral flow of current to the cathode short without gate assist results in forward biasing the edge of the cathode that is remote from the short. Use of the gate assist feature results in a countercurrent flow from the short to the gate, which reduces this lateral voltage drop. For any given value of gate-assisted current, a greater counteracting effect can be obtained if the resistance of the $P2$ base is higher. In addition, the use of a high resistivity $P2$ region results in reducing the gate assist power that is dissipated because of the ohmic current flow from the short.

In summary, therefore, a powerful technique for reducing the recovery time consists of reverse biasing the gate with respect to the cathode during reapplication of forward anode voltage. Shorted structures are not commonly used because of the difficulty of establishing values of reverse bias that are high enough for this application. Nevertheless, cathode shorts can be incorporated if the device is specifically designed with this in mind, and the resulting devices have improved performance capability.

The amplifying gate thyristor presents a special problem for gate-assisted turn-off because the negative gate bias can only divert the forward recovery current from the pilot thyristor. An approach to solving this problem is to place a rectifier diode across its gate-cathode terminals, to allow the diversion of forward recovery current from the main device. This is illustrated in the circuit configuration of Fig. 5.24. High voltage devices (1200 V) embodying this principle have been fabricated [36] with an integrated rectifier diode, to result in a fast recovery thyristor (6 μsec) with the short turn-on time associated with an amplifying gate structure (800 A/μsec). Although designed without cathode shorts, such devices exhibit a high dv/dt capability (200 V/μsec) during the application of a negative gate bias.

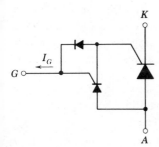

Fig. 5.24 Amplifying gate thyristor with gate-assisted turn-off.

5.5.1.3   The Reverse Conducting Thyristor

Another approach to obtaining fast recovery characteristics (which can be combined with lifetime reduction techniques) is to use device structures in which minority carriers can be swept out of the mid-region. The reverse conducting thyristor structure of Fig. 5.14, with both anode and cathode shorts, has been proposed as one solution to this problem [6], since both its base regions are accessible to the device terminals. The dc characteristics of such a structure are discussed in Section 5.3.3, where it is noted that the forward blocking capability of the RCT is independent of transistor action. As a consequence, this device can be [37] more heavily gold doped than conventional thyristors without deteriorating this important parameter.

It was also noted that the RCT does not have any reverse blocking capability. This characteristic causes no difficulties in many inverter systems, where feedback rectifiers are commonly wired in antiparallel with the thyristor for improved circuit operation. In fact, it would appear at first glance that the device configuration of Fig. 5.25a behaves like a conventional thyristor with a number of integrated antiparallel rectifier diodes ($P2N1N^+$), as a result of the anode and cathode shorts.

In actual practice, however, this presents a serious problem during turn-off, since this diode becomes forward biased when the voltage across the device is reversed, flooding the base regions with minority carriers. Thus the device will fail to achieve a forward blocking state upon reapplication of the supply voltage unless a long interval of time is provided during which no anode-to-cathode voltage is present, to permit these carriers to decay by recombination. A solution to the problem [38] is twofold. First, the $P1$ and $N2$ regions can be misaligned, to present a long forward path for these diodes, thus deteriorating their conductance characteristics. Second, a separate high conductance diode must be integrated into the structure, at some distance from the main thyristor (a few diffusion lengths), causing it to divert the flow of injected carriers during the supply reversal phase. Figure 5.25b is a physical realization for this structure.

Alternate approaches also considered [39] involve the use of separate contacts to the $N1$ and $P2$ regions. In these devices rapid turn-off is achieved by applying reverse base currents during the recovery phase. Perhaps the major disadvantage of these approaches is the difficulty of providing external connections to these regions in a high power device. In contrast, the reverse conducting thyristor principle employs internal contacts to these regions, which are made by conventional processing techniques, and avoids the attachment of additional leads.

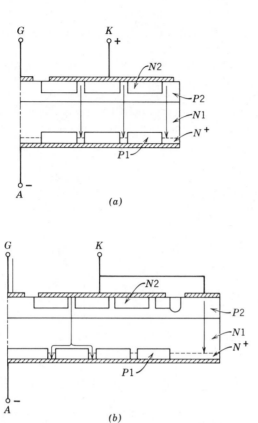

Fig. 5.25   The reverse conducting thyristor with intergrated diode.

5.6   GATE TURN–OFF

We have shown that the gate of a thyristor loses control once the device has been turned on; then no gate current is required to maintain it in its ON state. Under certain circumstances, however, it is possible to turn off a thyristor by the application of a reverse gate current. This gate turn-off capability is advantageous because it provides increased flexibility in circuit applications, where it now becomes possible to control power in dc circuits without the use of elaborate commutation circuitry. In principle, all thyristors can be turned off by a reverse gate bias. However devices designed specifically for this application are known as gate-controlled switches or gate turn-off thyristors. Such devices are often used in preference to transistors in high speed, high power applications, because of their ability to withstand higher voltages in their OFF state.

A one-dimensional description of the turn-off mechanism can be obtained by considering a p-n-p-n thyristor with a *reverse* gate current of magnitude I_G. All leakage terms are ignored in this analysis. The base drive required to sustain the $N2P2N1$ transistor in its ON state is equal to $(1 - \alpha_{npn})I_K$. The actual base current, however, is equal to $\alpha_{pnp}I_A - I_G$. Consequently turn-off will occur if

$$\alpha_{pnp}I_A - I_G \leqslant (1 - \alpha_{npn})I_K \qquad (5.40)$$

that is, if

$$I_G \geqslant \left(\frac{\alpha_{pnp} + \alpha_{npn} - 1}{\alpha_{npn}} \right)I_A \qquad (5.41)$$

since $I_A = I_G + I_K$. Defining I_A / I_G as the turn-off gain, β_{off}, gives

$$\beta_{off} \leqslant \left(\frac{\alpha_{npn}}{\alpha_{pnp} + \alpha_{npn} - 1} \right) \qquad (5.42)$$

Typical values for turn-off gain range between 5 and 10.

A high turn-off gain can be obtained by making the alpha of the n-p-n as close to unity as possible, by designing it with a narrow base ($P2$) and a heavily doped emitter ($N2$) region. At the same time, the alpha of the p-n-p must be kept small by making its ($N1$) base region wide. Additionally, gold doping or some other lifetime control technique is used for this purpose. As a consequence, these devices usually have fast recovery characteristics and a relatively large voltage drop in the ON state, compared to conventional thyristors.

5.6.1 The Squeezing Velocity

The arguments just presented are somewhat simplistic because they assume that both the n-p-n and the p-n-p transistors are on the edge of saturation prior to turn-off. Actually, since both devices are heavily saturated in their ON state, the removal of excess stored charge is an important part of the turn-off process. This results in a storage time delay, followed by a fall time during which the device returns to its nonconducting state.

Stored charge is directly removed from $P2$ by the gate current. This removal is essentially a two-dimensional process and is the inverse of the spreading that occurs in this region during turn-on. This is because of the voltage drop due to the lateral flow of gate current in $P2$, which results in the J_3 junction becoming less positively biased as we proceed from the center of the device toward the gate contact. Eventually a part of J_3 that is closest to the gate contact becomes reverse biased. At this point, therefore, all the forward current will be squeezed through the remaining part of J_3 that is still forward biased, resulting in an increase in the local emitter current density. This squeezing action is analogous to emitter crowding with two important differences. First, all the current crowding occurs under the emitter region that is *farthest* from the gate contact. Second, squeezing is a dynamic process, whereas emitter crowding in a transistor is a steady state phenomenon. Here the forward current is progressively squeezed into a smaller and smaller emitting region, until some limiting dimension δ is reached.* At this point, the remaining excess charge in $P2$ is removed, and the storage phase is over.

To an approximation, the conducting region can be considered to consist of a space charge neutral plasma. The dynamics of the squeezing process can be determined by evaluating [40] the rate of removal of charge from the edge of this plasma at some point x_b and over an element of thickness Δx_b (see Fig. 5.26a, featuring a symmetrical structure of thickness Z). Recombination is ignored within this volume element. On the left-hand side, charge is added by the flow of a lateral hole diffusion current $I_x(x_b)$ from the ON region. At the same time, charge is extracted from the right-hand side to provide the gate current $I_G/2$. If ΔQ is the hole charge in this volume, then

$$\frac{\Delta Q}{\Delta t} = I_x(x_b) - \frac{I_G}{2} \tag{5.43}$$

This equation assumes that no excess charge enters the element in the z direction and that all the gate current is due to the extraction of holes from the volume element. These are relatively poor assumptions, since J_2 may either be forward or reverse biased during this process. However it serves as a starting point and can supply some insight into the squeezing process. The respective terms in (5.43) may be calculated by making additional assumptions about the variation of minority carrier concentration in the $P2$ base region. The electron concentrations in the base can be written as

$$n(x,y) = n(y)f(x) \tag{5.44}$$

*This is necessary—otherwise the local emitter current density would become infinite just before the end of the storage phase.

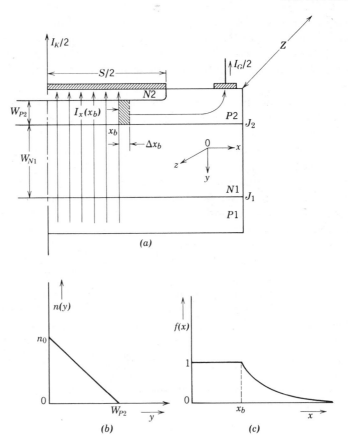

Fig. 5.26   Current flow in a gate turn-off thyristor.

where

$$n(y) = n_0 \left(1 - \frac{y}{W_{P2}}\right) \tag{5.45}$$

This is a reasonable assumption, since $N2P2N1$ operates as a narrow base transistor (Fig. 5.26b). The electron concentration in the x direction is assumed constant up to x_b, at the edge of the moving plasma boundary, falling off exponentially with distance beyond this point. Thus, as shown in Fig. 5.26c,

$$f(x) = 1 \qquad\qquad 0 \leqslant x \leqslant x_b \tag{5.46a}$$

$$f(x) = \exp -\left(\frac{x - x_b}{x_0}\right) \qquad x_b \leqslant x < \infty \tag{5.46b}$$

The constant x_0 is on the order of an electron diffusion length and is actually a complex function of the electric field, the high level lifetime, and the ambipolar parameters. The limiting width of the plasma, prior to extinction, is on the order of this quantity.

The excess hole charge in the elemental volume, of thickness Δx_b, is given by

$$\Delta Q = \Delta x_b \int_0^{W_{P2}} Zqn(x,y)\,dy \qquad (5.47)$$

where Z is the depth dimension, provided $f(x) \cong f(x_b) = 1$ throughout this narrow region. Solving, we have

$$\Delta Q = \frac{qZn_0 W_{P2}\Delta x_b}{2} \qquad (5.48)$$

The hole current that flows into the element is given by

$$I_x(x_b) = -\int_0^{W_{P2}} ZJ_x(x_b,y)\,dy \qquad (5.49)$$

where

$$J_x(x_b,y) \cong qD_n \frac{\partial n(x,y)}{\partial x}\bigg|_{x=x_b} \qquad (5.50)$$

Combining (5.44) with (5.46), (5.49), and (5.50) gives

$$I_x(x_b) = \frac{qn_0 ZD_n W_{P2}}{2x_0} \qquad (5.51)$$

The term n_0 in this equation is evaluated by noting that

$$J_y(x,y) = qD_n \frac{\partial n(x,y)}{dy} \qquad (5.52)$$

Consequently the cathode current, in the direction shown in Fig. 5.26a, is given by

$$I_K/2 \cong -Z\int_0^\infty J_y(x,y)\,dx \qquad (5.53a)$$

Substituting into (5.52) and (5.44) to (5.46) gives

$$I_K/2 = \frac{qn_0 ZD_n(x_b + x_0)}{W_{P2}} \tag{5.53b}$$

Equation 5.43 can now be solved with the aid of (5.48), (5.51), and (5.53), resulting in

$$\frac{\Delta x_b}{\Delta t} = -\frac{I_G}{I_K} \frac{x_b + x_0}{t_{P2}} + \frac{D_n}{x_0} \tag{5.54}$$

where t_{P2} is the transit time through the gated base. Since $\beta_{\text{off}} = I_A/I_G$, this can be modified to

$$\frac{\Delta x_b}{\Delta t} = -\frac{x_b + x_0}{t_{P2}(\beta_{\text{off}} - 1)} + \frac{D_n}{x_0} \tag{5.55}$$

This is the velocity of movement of the ON state boundary.

The time taken for the plasma to be squeezed from $S/2$ to δ is the storage time t_s, which is given by integrating (5.55) over these limits. Solving, and noting that $\delta \ll S/2$, we write

$$t_s = t_{P2}(\beta_{\text{off}} - 1)\ln\left(\frac{Sx_0/W_{P2}^2 + 2x_0^2/W_{P2}^2 - \beta_{\text{off}} + 1}{4x_0^2/W_{P2}^2 - \beta_{\text{off}} + 1}\right) \tag{5.56}$$

Thus the storage time increases with increasing values of turn-off gain.

5.6.2 Turn-off Gain

From (5.56) it is seen that

$$\beta_{\text{off}}(\text{max}) = 1 + \frac{4x_0^2}{W_{P2}^2} \tag{5.57}$$

at which point the storage time becomes infinite. This result is quite different from (5.42), which was obtained for the one-dimensional case. In effect, we can say that (5.57) gives the maximum gain that can be used to take the thyristor out of saturation, whereas the gain given by (5.42) is the maximum that can be used to turn off a thyristor which is just on the edge of saturation. The limiting gain is the smaller of the two.

5.6.3 Maximum Anode Current

As mentioned earlier, a reverse voltage builds up across J_3 because of the lateral flow of current through the gated base. The upper limit on the gate current, hence on the turn-off time, is reached when this junction breaks down.

The maximum reverse voltage across J_3 occurs when the entire plasma is squeezed to its limiting width of $\delta \ll S/2$. For this case, it is seen that

$$I_G \leqslant \frac{4}{R_{P2}} BV_{J3} \qquad (5.58)$$

where R_{P2} is the total resistance of the $P2$ base as measured between gate contacts, and BV_{J3} is the breakdown voltage of the J_3 junction. It follows from (5.58) that the maximum anode current that can be turned off is given by

$$I_A \leqslant 4\frac{\beta_{\text{off}}}{R_{P2}} BV_{J3} \qquad (5.59)$$

5.6.4 The Fall Time

The bias over the entire J_2 junction changes from forward to reverse at the end of the storage phase. This is followed by a time interval known as the fall time, during which the forward current is reduced to zero and full blocking voltage is established across J_2. Simultaneously, the anode-cathode voltage is increasing during this process; therefore appreciable power is being dissipated in the thyristor during this phase.

The dominant process [41] during the fall time is the expansion of the depletion layer that builds up across J_2. This is essentially in the $N1$ base and results in a sweeping out of the hole charge in this wide region. The sweeping-out process may be analyzed on the assumption that the hole concentration can be represented by an average value during this phase. Figure 5.27 depicts the electrical circuit that must also be considered in this analysis. With reference to this figure,

$$I(t) \cong qAp^* \frac{dx}{dt} \qquad (5.60)$$

where p^* is the average hole concentration in the $N1$ base. Assuming that

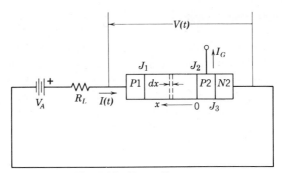

Fig. 5.27   Turn-off process.

the $P2N2$ junction is an abrupt one, the depletion layer width is given by

$$x = k_v (V)^{1/2} \tag{5.61}$$

where k_v depends on the doping levels of $P2$ and $N2$, and V is the voltage drop across J_2. Finally, the circuit equation is

$$V_A - V(t) = I(t)R_L \tag{5.62}$$

where R_L is the load resistance and voltage drops across J_1 and J_3 are ignored.

Combining (5.60) with (5.62) and solving, gives

$$I(t) = I_A \operatorname{sech}^2 \left[\frac{I_A t}{qAk_v p^* (V_A)^{1/2}} \right] \tag{5.63}$$

Thus the current through the device falls from its initial value of I_A to 10% of this value in a time given by

$$t_f = \frac{1.82 q A k_v p^* (V_A)^{1/2}}{I_A} \tag{5.64}$$

To an approximation, therefore, the fall time increases with increasing anode voltage when the anode current is kept constant, but it decreases with increasing anode current for constant anode voltage. These results have been verified in practice.

5.6.5 Concluding Remarks

Reliable operation of gate turn-off thyristors can be achieved only if the final area into which the plasma is squeezed is uniformly distributed and large enough to prevent destruction of the device. Thus uniformity of the cathode-gate spacing is essential. This requirement has resulted in the use [21, 42] of highly interdigitated configurations such as the involute structure of Fig. 5.19, which calls for a large value of gate drive because of its long gate-cathode perimeter. Consequently the use of an amplifying gate is highly desirable, to be able to obtain fast turn-on. This, in turn, necessitates the incorporation of a rectifier diode in shunt with the pilot thyristor, to turn off the main thyristor.

It is seen, therefore, that the gate turn-off thyristor and the gate-assisted turn-off thyristor share many features. Thus both devices are often designed without cathode shorts because of the lateral flow of current from the short to the reverse-biased gate. Both devices use highly interdigitated cathode-gate structures and have long injecting perimeters. Both devices often incorporate amplifying gates to handle the large gating currents needed for fast turn-on, and both require the use of rectifier diodes as shown in Fig. 5.24. The essential difference is that the gate turn-off thyristor is brought out of conduction by the application of a negative bias on the gate, while the anode is positive with respect to the cathode. On the other hand, the gate-assisted turn-off thyristor requires commutation of the supply voltage to turn it off, and a reverse gate bias to improve its recovery characteristics.

5.7 BIDIRECTIONAL THYRISTORS

Up to this point we have focused attention on devices that operate from dc power supplies. On the other hand, the majority of applications are based on the control of energy that is derived from the power lines. Thus there is a great need for devices that can be operated directly from an ac power source, since they can greatly simplify the associated circuitry and reduce both the size and cost of power control equipment.

A number of different bidirectional structures have been developed for these applications. In general, they behave like two conventional thyristors connected in antiparallel, to permit the accommodation of signals of both polarities, as in Fig. 5.28a. The integration of this arrangement into a single two-terminal device can be accomplished only by making use of the shorted-cathode principle. Thus the use of shorts can make available new device configurations that are otherwise not possible [5]. Figure 5.28b

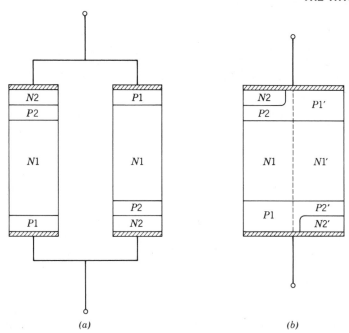

Fig. 5.28 Bidirectional thyristors.

shows a structure, which can be formed by diffusing masked symmetrical n^+-regions into both sides of a previously diffused *p-n-p* slice. The inherent symmetry of this structure results in identical performance for either polarity of applied signal.

Although no control gate is incorporated into this device, it is useful in a number of applications in which it can be triggered into conduction by exceeding the forward breakover voltage (breakover triggering). Alternately, triggering can be accomplished by exceeding the rate of rise of forward voltage (dv/dt triggering) across its terminals.

The problem of providing gate control in these structures is considerably more complex, since a single gate region must serve for both the antiparallel thyristors. From practical considerations, this gate must be located on one side of the device, which successively adopts the roles of anode and cathode during operation. In effect then, the gate must be capable of controlling the device from either its "anode" side or its "cathode" side. The strategy here is to exploit a number of physical mechanisms for controlling a thyristor and to incorporate them into a single unit [43]. In this way, a single gate can be made to provide control on both half-cycles if it can operate in a number of ways, depending on the polarity of the voltage across its main terminals and on the direction of flow of gate current. Two such gate control methods are now considered; the shorted structure is necessary for device operation in both.

5.7.1 Junction Gate Control

Figure 5.29a shows a conventional thyristor, $P1N1P2N2$, with an additional $N3$ region to which the gate lead is connected. The device is biased with main terminal ② positive with respect to main terminal ①, so that ① services as the cathode for this half-cycle of power supply. It is now shown that this device can be turned ON for either direction of gate current flow.

Consider the thyristor in its blocking state, where J_1 and J_3 are forward biased, and J_2 is reverse biased. For convenience, the device can be thought of as comprising a "main" thyristor, and an "auxiliary" thyristor in physical contact. The lateral resistance of $P2$ is relatively low because of its heavy doping. Consequently the devices are strongly coupled by this base region. On the other hand, since $N1$ is lightly doped, coupling by way of this base is relatively slight and is ignored. Its inclusion, however, results in minor modifications in describing the thyristor action.

A gate signal is now applied in the direction indicated in Fig. 5.29a, by making the gate negative with respect to ①. This causes J_3 to be reverse biased, preventing the main thyristor from operating. At the same time, J_4 is forward biased and provided with a flow of current I_1 into its base. Note that the presence of the short on $N2$ is necessary; on the other hand, the short on $N3$ is not, but it does not materially alter the situation.

The base current I_1 serves to turn ON the auxiliary $P1N1P2N3$ thyristor when the sum of its localized alphas exceeds unity. Current flow through this thyristor continues to build up until the gate resistance limits the injection of electrons into $P2$. On the other hand, the hole supply from terminal ② is essentially unlimited. For charge neutrality to be preserved in the $P2$ region, additional electrons must enter it by back injection across J_2 (i.e., this junction becomes locally forward biased). In addition, excess holes in $P2$ are removed by supplying a gate current I_2 into the main thyristor $P1N1P2N2$ (Fig. 5.29b). This causes J_3 to become forward biased so that the main thyristor now turns ON, with initial conduction occurring at the edge closest to the gate. Conduction spreads until the entire J_2 junction becomes forward biased, and the main thyristor turns fully ON.

In summary, therefore, a negative gate current triggers the auxiliary thyristor into forward conduction. As this device turns ON, it provides gate drive for the main thyristor, which starts to conduct at the edge that is

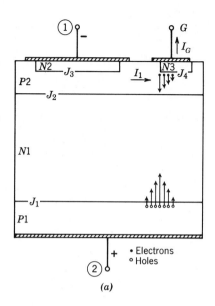

(a)

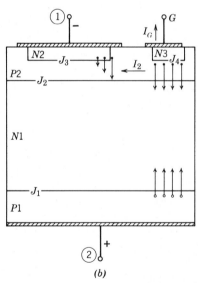

(b)

Fig. 5.29 Junction gate operation.

closest to the gate. This conduction propagates over the entire $N2$ region as the device becomes fully turned ON.

Consider now the half-cycle during which the gate terminal is made positive with respect to ①. Operation of the device is now relatively straightforward: J_4 becomes reverse biased, but gate current flows *into* $P2$ through the short, resulting in turn-on of the main device by conventional thyristor action.

The use of junction gate triggering thus allows gate signals of *either* polarity to be used for triggering the thyristor when its main terminal ① serves as its cathode and main terminal ② serves as its anode. Bipolar gate control is accomplished from the "cathode" side of the thyristor in this manner.

5.7.2 Remote Gate Control

Remote gate control allows a thyristor to be triggered when main terminal ① is positive with respect to main terminal ② (i.e., from its "anode" side). This structure is identical to that of Fig. 5.29, except for the inclusion of an additional n-diffusion ($N4$) into its "cathode" side, where terminal ② is located (see Fig. 5.30a). Now the main thyristor consists of $P2N1P1N4$, with $P2$ as its anode. During its full conduction, the region $N2$ will be inoperative because of the short to the $P2$ region.

Consider this thyristor in its forward blocking state, with J_2 and J_5 forward biased and J_1 reverse biased. At this point the forward current, flowing from ① to ②, is given by the leakage current, which is negligible for silicon devices. A gate current is now applied in the direction shown, by making the gate terminal negative with respect to terminal ①. This causes current I_1 to flow, and junction J_4 becomes forward biased, injecting electrons that diffuse across P_2 and are subsequently collected at $N1$. Note that the direction of the electric field across this junction is such that electron collection will occur at $N1$, even though J_2 is forward biased. In effect, $N3P2N1$ behaves like a saturated transistor with $N3$ as its emitter. The collected electrons lower the potential of $N1$ with respect to $P2$, allowing more forward current to flow across the entire junction J_2, eventually triggering the main thyristor into conduction.

The use of a short on the gate does not materially alter the operation of the device. However the short on main terminal ① allows the remote gate principle to be extended to the situation of the gate being made positive with respect to the anode. Figure 5.30b illustrates this situation, when ① is positive with respect to ② as before. Now, however, the gate terminal is made positive with respect to the anode, and the junction J_3 becomes

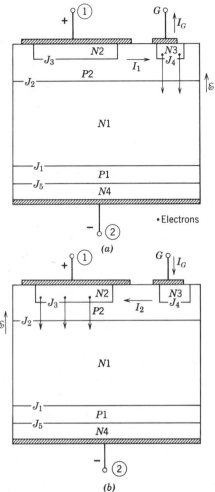

Fig. 5.30 Remote gate operation.

forward biased and injects electrons that are collected at $N1$, eventually triggering the main thyristor $P2N1P1N4$ into conduction. Once this occurs, however, the $N2$ region plays no further role, since the main thyristor current will be carried through the shorted part of the anode that is in contact with $P2$. The short across J_4 is important for this mode of triggering, since it allows J_3 to become forward biased during the triggering process (when the gate is made positive with respect to main terminal ①).

In summary, then, it is possible to trigger a thyristor into conduction from its anode side by means of remote gate control. The use of two shorted diffusions on the anode side allows triggering to be achieved by gate signals of either polarity with respect to the anode.

5.7.3 The Triac

A number of bidirectional structures can be designed using the principle just outlined. Perhaps the most useful one, however, is a structure that exploits the combination of junction gate control, remote gate control, and conventional thyristor action, resulting in a bidirectional triode switch that is capable of direct operation from the ac power supply. Such a device, known as a triac, can be triggered into conduction by gate signals of either polarity. It provides an extremely flexible means for power control because it eliminates many of the auxiliary components normally used in ac circuits with conventional thyristors.

Figures 5.31a and 5.31b provide different views of this device. Device operation can be more easily followed with reference to Fig. 5.31c, which is a two-dimensional representation. Main device terminals, indicated as ① and ②, alternately share the roles of anode and cathode during successive half-cycles of applied voltage.* Note that this device is identical to that of Fig. 5.30, except that $N4$ and $P1$ are electrically connected by means of a short at main terminal ②.

Note that the structure is inherently symmetrical except for the gate region and its associated metallization. Conventionally, this gate is placed on the ① face of the device; since its area is small compared to the rest of the structure, the lack of symmetry created by its presence is relatively slight.

Device behavior is that of a symmetrical bidirectional diode switch if the gate lead is left open. Such a device (Fig. 5.28b) can be electrically triggered into conduction by either dv/dt or forward breakover techniques, as described earlier. Four combinations of terminal polarity and gate current directions are possible when the device is operated as a triode. Each of these is now considered separately, with reference to Fig. 5.31c.

1. Terminal ① positive with respect to ②; I_G negative. The device behaves like a thyristor with remote gate control on its anode side (see Section 5.7.2). In this mode of operation, triggering is initiated by injection of electrons from the $N3$ region into the $N1$ region. This lowers the potential at $N1$, causing holes to be injected from $P2$ into this region. These holes provide the base current drive for the $P2N1P1$

---

*These terminals are sometimes referred to in the literature as the "cathode" and "anode," respectively.

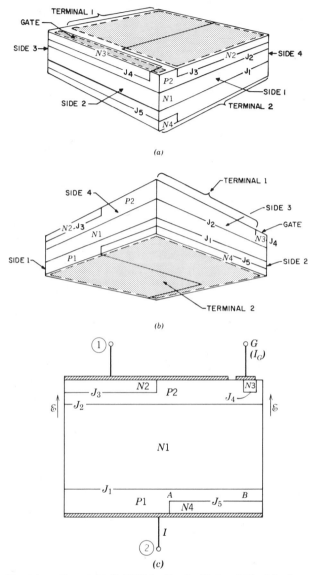

Fig. 5.31 The triac. Copyright © 1965 by the Institute of Electrical and Electronics Engineers, Inc. Reprinted with permission from Gentry et al. [43].

transistor, and the $P2N1P1N4$ device eventually triggers into conduction. During operation, J_3 is reverse biased; thus the main current I is carried through the short at terminal ①. On the other hand, full device current is carried through the $N4$ region, not through the short at terminal ②.

2. Terminal ① positive with respect to ②; I_G positive. In this situation device triggering is again by remote gate control. The junction J_3 becomes forward biased, resulting in injection of electrons from $N2$ into $N1$, accompanied by increased forward bias of the junction J_2. Eventually full current is carried through the short at ①. The gate junction J_4 is reverse biased and is inactive in this situation. As before, full device current is carried through the $N4$ region, not through the short at terminal ②.

3. Terminal ① negative with respect to ②; I_G positive. The device now behaves as a conventional thyristor, with gate current supplied through the short on the $N3$ region. Since region $N4$ is inactive during conduction, current is now carried through the short on the terminal ② side. On the other hand, $N2$ is an active region; therefore it carries the main current at terminal ①.

4. Terminal ① negative with respect to ②; I_G negative: This situation results in device operation by junction gate control, as described in Section 5.7.1. Turn-on is a two-step process, with initial gating of the auxiliary thyristor $P1N1P2N3$ by the flow of lateral base current in $P2$, toward the $N3$ gate. Full conduction of this thyristor results in a flow of base current *out* of this device and toward the $N2$ region. This provides gate current for the main thyristor $P1N1P2N2$, which is then triggered into conduction.

Triggering of the main device can occur in an alternate manner, depending on the thickness of the $P1$ region that lies between $N4$ and $N1$, and on its sheet resistance. As before, application of a negative gate current drives lateral base current into the auxiliary thyristor, so that $P1N1P2N3$ begins to conduct. This results in a lateral voltage drop from A to B (Fig. 5.31c) in region $P1$, causing point A to become more forward biased. This can result in the injection of holes into $N1$ and conduction of the $P1N1P2N2$ thyristor, before the local thyristor is fully ON. In either event, however, full conduction of the main thyristor is accomplished in this mode of operation, with the main current being carried through the short at terminal ② and through $N2$ at terminal ①.

Figure 5.32a shows the voltage-current characteristics across the main terminals (i.e., the output characteristics of this device) for all four modes of device operation. Behavior is almost perfectly symmetrical, as expected,

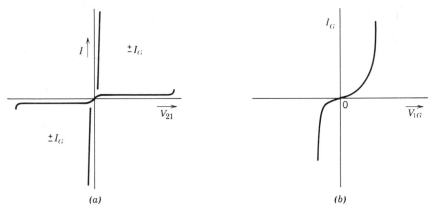

Fig. 5.32   Terminal characteristics of a triac. (a) Output, (b) Gate.

since the structure is made by simultaneous diffusions on either side of the silicon slice. Lack of device symmetry is observed in the gate input characteristic, which shows the gate current as a function of the voltage across the gate and main terminal ①. Inspection of Fig. 5.31c reveals that the gate input consists of two highly dissimilar junction diodes, connected in antiparallel and shunted with a common resistance.* A typical input characteristic of this type appears in Fig. 5.32b.

The gate current required to trigger the triac into conduction depends on the specific mode of operation. Thus for situations involving remote gate control, the case of I_G positive will result in a higher value of gating current, since $N2$ behaves as the emitter instead of the much smaller $N3$ region. Gating currents for the junction gate and conventional thyristor operation modes both result from the lateral flow of current in $P2$ and tend to be relatively similar in magnitude, since both are a function of the sheet resistance of the $P2$ region. For all situations, however, the gate current is relatively independent of the supply voltage, as well as of the load current. This important lack of interaction between input and output greatly simplifies the design of triac circuits.

In summary, the triac is a symmetrical triode switch, capable of controlling loads that are supplied with ac power. Control is effected by the introduction of either positive or negative gating currents. Thus control signals can be derived from the power supply as well, resulting in many circuit simplifications. The device relies for its operation on the principles of junction gate control, remote gate control, and conventional thyristor

---

*Often a groove is etched in the $P2$ region, between the gate and main terminal ①, to increase the value of this shunting resistance.

action, and on the extensive use of shorts, which cover almost half its surface.

The integration of what is essentially two devices on a single piece of silicon results in only half the structure being used at any one time. Consequently triac area utilization is poor and is about equal to that of two independently connected thyristors. The main advantages of fabrication in a single chip are the almost perfect matching of output characteristics, which would be possible only by careful preselection of individual devices. In addition, the elimination of one package, as well as a number of external connections, results in substantial economic advantages for this type of device. Finally, the incorporation of techniques that allow bidirectional control signals results in much more convenience to the user, with little extra cost to the manufacturer. It is not surprising, therefore, that the development of these devices has rapidly encompassed a wide range of operating voltages (up to 1400 V) and currents (in excess of 200 A).

5.7.3.1 Commutating dv/dt

As we have indicated, the triac is primarily used with ac power supplies. In a typical circuit application, conduction during one half-cycle is immediately followed by the reapplication of a half-cycle of opposite polarity. Satisfactory operation requires that the triac be capable of blocking at this point, to permit gate control to be regained.

The triac can be modeled as two antiparallel thyristors, one operating on positive half-cycles and the other on negative half-cycles of the same power supply. If separated, each thyristor will have some charge left in its bases at the end of its conduction period. The presence of this charge can result in failure to remain OFF when the supply voltage is reapplied. In a triac, moreover, these devices are physically connected; thus charge can be laterally transported from the conducting device over into the bases of the device which must block. As a result, the allowable rate of rise of reapplied voltage, or commutating dv/dt, can be greatly reduced from what it would be for physically separated devices.

Section 5.5 noted that the forward recovery current of a thyristor plays a key role in determining the ability of the device to block the reapplication of anode voltage. Furthermore, this recovery current increases with the rate of removal of the anode current. This is also seen in a triac [44], where the commutating dv/dt has been found to be approximately inversely proportional to this removal rate.

The increase of carrier lifetime with temperature also adversely affects the allowable dv/dt of a conventional thyristor. Again, the overall effect on the triac is stronger than on the conventional thyristor, because of the interaction of the antiparallel devices. Typically the commutating dv/dt of a triac has been found to fall by a factor of 2 for every 40°C rise in temperature.

The problems of low commutating dv/dt has severely limited the ease with which this device can be used in circuits other than those operating at 60–400 Hz. Even here, inductive loads can result in excessively high values of anode current removal rate, accompanied by fast rates of rise of reapplied voltage. Circuit techniques frequently are used to control these fast signals. From a device point of view, interaction can be reduced by physically arranging the device to ensure maximum separation between its individual sections. Different alignment arrangements of the n-regions ($N2$ and $N4$ in Fig. 5.30) have been used to obtain an improvement in this parameter, with some success.

5.7.4 Trigger Devices

The inherent simplicity of the triac makes it ideal for use when overall circuit costs must be kept to a minimum. Consequently a series of trigger devices has been developed to compensate for the fact that its firing voltage may vary from 0.5 to 3 V because of variations in temperature and variations from one device to another.

A trigger device must have two stable states—a high impedance state that limits the flow of current to well below the firing level, and a low impedance state that passes ample current to fire the triac under worst case conditions. In addition, triggering must occur at a low voltage that is relatively invariant with temperature, or from device to device.

The bidirectional thyristor diode, sometimes called the diac, is most commonly used for this application (see Fig. 5.28b). Typically, this device is designed to have a breakover voltage of between 25 and 40 V, and it is suited for applications in which gate drive is also obtained from the power line. The threshold voltage of such a device is governed by all the considerations described in Section 5.2. However the depletion layer width of the blocking junction is relatively small in this case, which means that changes in leakage current with temperature are of little consequence. This fact, coupled with the use of a shorted emitter structure, makes these devices relatively insensitive to temperature variations.

Some applications call for low, relatively precise values of trigger voltage. Here, a common approach is based on the following properties: (a) the breakdown voltage of the emitter-base diode of a conventional diffused transistor is low (6–7 V for a junction depth of 2–2.5 μm), and (b) it is almost insensitive to temperature variations. Such a diode can be used to provide gating current to a p-n-p-n thyristor, which can thus abruptly switch into conduction at this value of BV. Two such diode-thyristor combinations, placed in antiparallel, can be used to form a bidirectional trigger, whose circuit diagram appears in Fig. 5.33a. Note that both thyristors are triggered at the $N1$ base by removal of current when the avalanche breakdown voltage of the diode is exceeded.

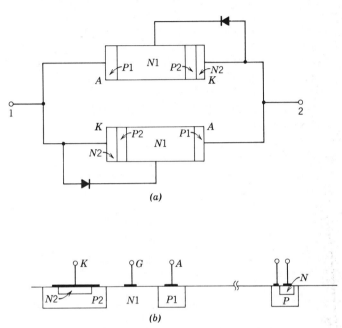

(a)

(b)

Fig. 5.33 Details of a bidirectional switch.

Conventional integrated circuit techniques can be used for fabricating this bilateral trigger circuit [45]. Figure 5.33b illustrates how a low breakdown voltage diode and a lateral thyristor [46] can be made on the same chip by compatible monolithic processes. Note the use of a shorted cathode structure to provide an improved dv/dt capability for the thyristors. Alternately, diffused resistors can be placed across the $P2N2$ regions to provide the same function [47]. Interaction between the thyristors can be minimized if the two devices are physically spaced a few minority carrier diffusion lengths apart. Conventional junction isolation techniques can also be used, but this leads to a considerable increase in device process complexity. Finally, either of these bilateral switches can be made in unilateral form, for use as triggering elements with conventional thyristors.

5.8 INSTABILITIES AND FAILURE MODES

There are a number of ways in which a thyristor can fail to meet the specifications for which it was designed. For example, the reverse blocking voltage can fall below its designed value if there are local inhomogeneities in its depletion layer. The power dissipation in the ON state can be excessive if the device has an unduly short lifetime because of poor processing. Improper mounting of thyristors, especially devices that are soft soldered to the case, often results in the formation of voids that create localized regions of high thermal resistance. Device failure is often found to occur because of the localized concentration of heat at these voids. All these types of failure are generally cause for rejection of the device; yet they can be minimized by careful processing techniques, some of which are described in Chapter 6. This section considers only the instabilities and failure mechanisms that arise during the operation of an otherwise acceptable device that has met its rating specifications.

A number of device failure modes are more correctly called malfunctions. In some instances they can result in device failure, depending on the circuit application. On the other hand, some failure modes create the buildup of power in highly localized regions of the device, invariably leading to its destruction by mesoplasma formation and second breakdown. One approach to understanding the different failure modes is with reference to the two-transistor analogy for a thyristor, which has been found to be useful in describing device action up to the initiation of forward conduction. This analogy (Fig. 5.34) models the thyristor as a two-stage, positive feedback amplifier, consisting of a p-n-p transistor and an n-p-n transistor.

The terminal characteristic of this amplifier is represented in Fig. 5.35, where the region 0–A depicts one static state (forward blocking) and B–C depicts the other (forward conduction). Both these states are inherently stable and exhibit a positive dynamic resistance. Ordinarily the thyristor is operated as a switch, spending most of its time in one of these static, or stable, states. Each of its equivalent transistors is operating at very low current levels during forward blocking; during forward conduction, both are saturated. In either event, the forward gain in the feedback loop is negligible.

The region A–B depicts the transition from blocking to conduction and is initiated when the sum of the ac alphas of the transistors exceeds unity. Over this region, both its equivalent transistors are operating in their active, high gain regime, rendering the circuit of Fig. 5.34c unstable. In fact, it can even be made to oscillate if inductively terminated, either by the load or by improper wiring.

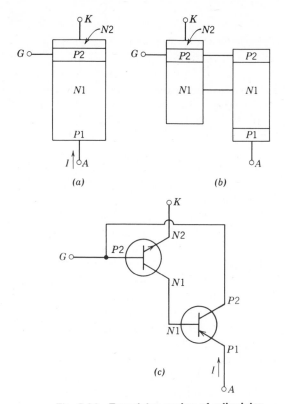

Fig. 5.34 Transistor analog of a thyristor.

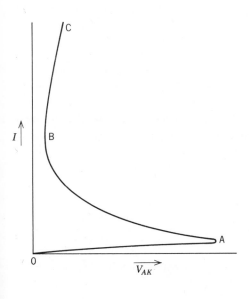

Fig. 5.35 Output characteristics of
a thyristor (expanded scale).

257

The terminal characteristics of Fig. 5.35 indicate that the negative resistance is the short-circuit-stable type, for which current flow occurs by filamentation. In practice, the region A–B represents an extremely small change in forward current. Nevertheless any turn-on process that is not rapid can cause the device to dwell in this region, ultimately leading to its destructive failure by mesoplasma formation and second breakdown. This is often observed in two-terminal thyristors [48] that are switched by increasing the anode-cathode voltage (breakover triggering). Here uniform turn-on is achieved only if the device is overdriven with a steep voltage ramp, which results in simultaneous dv/dt and breakover triggering, hastening the turn-on process.

The operation of a thyristor in its negative resistance region can lead to localized current flow, even if it is assumed that the device is ideally uniform to start with. Additional failure mechanisms occur, however, if the device has inhomogeneities such as those caused by crystal defects, damage, or precipitates. These "soft spots" can lead to local variations in the alphas of the p-n-p and n-p-n transistors associated with thyristor action. Thyristors that are fired by exceeding the forward blocking voltage (breakover triggering) are usually found to turn on first at such points within the semiconductor bulk. If the forward current is reduced to below the holding current, the device will turn off last at the same point. Thus the soft spot plays a significant role in localizing device action during these transitions. Again, failure in this manner can be alleviated by overdriving the thyristor.

Local inhomogeneities as well as crystal defects can also cause uneven turn-on of a thyristor. Thus the initiation of plasma in thyristors made on (111)-oriented silicon is often found to occur preferentially in the [211] direction [49]. Here too, a large gate drive makes the initial turn-on process more uniform and masks the effect of these inhomogeneities.

Device failure during the dynamic phase is a particularly severe problem with gate turn-off thyristors. In these devices the turn-off process consists of dynamically squeezing the plasma into a small region until the storage phase is terminated. These devices must be highly interdigitated, to make the final turn-off area as large as possible. In addition, every effort must be made to terminate the storage phase rapidly. This, in turn, results in the achievement of very low values of turn-off gain in practical situations.

5.9 REFERENCES

1. J. L. Moll et al., "*P-N-P-N* Transistor Switches," *Proc. IEEE*, **44**, No. 9, pp. 1174–1182 (1956).

2. F. E. Gentry et al., *Semiconductor Controlled Rectifiers*, Prentice-Hall, Englewood Cliffs, N.J., 1964.

3. A. Herlet, "The Maximum Blocking Capability of Silicon Thyristors," *Solid State Electron.*, **8**, No. 8, pp. 655–671 (1965).

4. W. Fulop, "Three Terminal Measurements of Current Amplification Factors of Controlled Rectifiers," *IEEE Trans. Electron Devices*, **ED-10**, No. 3, pp. 120–133 (1963).

5. R. W. Aldrich and N. Holonyak, Jr., "Two-Terminal Asymmetrical and Symmetrical Silicon Negative Resistance Switches," *J. Appl. Phys.*, **30**, No. 11, pp. 1819–1824 (1954).

6. R. A. Kokosa and B. R. Tuft, "A High-Voltage High-Temperature Reverse Conducting Thyristor," *IEEE Trans. Electron Devices*, **ED-17**, No. 9, pp. 667–672 (1970).

7. R. A. Kokosa, "The Potential and Carrier Distributions of a *p-n-p-n* Device in the ON State," *Proc. IEEE*, **55**, No. 8, pp. 1389–1400 (1967).

8. A. Herlet and K. Raithel, "Forward Characteristics of Thyristors in the Fired State," *Solid State Electron.*, **9**, No. 11/12, pp. 1089–1105 (1966).

9. E. S. Yang and N. C. Voulgaris, "On the Variation of Small Signal Alphas of a *p-n-p-n* Device with Current," *Solid State Electron.*, **10**, No. 7, pp. 641–648 (1967).

10. M. Kurata, "One-Dimensional Calculation of Thyristor Forward Voltages and Holding Currents," *Solid State Electron.*, **19**, No. 6, pp. 527–535 (1976).

11. J. Cornu and A. Jaecklin, "Processes at Turn-On of Thyristors," *Solid State Electron.*, **18**, No. 7/8, pp. 683–689 (1975).

12. F. Dannhauser and P. Voss, "A Quasi-Stationary Treatment of the Turn-On Delay Phase of One-Dimensional Thyristors: Part 1—Theory. Part 2—Experiment," *IEEE Trans. Electron Devices*, **ED-23**, No. 8, pp. 928–939 (1976).

13. N. Mapham, "The Ratings of Silicon Controlled Rectifiers when Switching into High Currents," *IEEE Trans. Commun. Electron.*, **83**, No. 74, pp. 515–519 (1964).

14. R. L. Longini and J. Melngailis, "Gated Turn-On of Four Layer Switch," *IEEE Trans. Electron Devices*, **ED-10**, No. 3, pp. 178–185 (1963).

15. H. J. Ruhl, Jr., "Spreading Velocity of the Active Area Boundary in a Thyristor," *IEEE Trans. Electron Devices*, **ED-17**, No. 9, pp. 672–680 (1970).

16. T. Matsuzawa, "Spreading Velocity of the ON State in High Speed Thyristors," *Electr. Eng. Japan*, **93**, No. 1, Part C, pp. 136–141 (1973).

17. W. H. Dodson and R. L. Longini, "Probed Determination of Turn-On Spread of Large Area Thyristors," *IEEE Trans. Electron Devices*, **ED-13**, No. 5, pp. 478–484 (1966).

18. Y. Yamasaki, "Experimental Observation of the Lateral Plasma Propagation in a Thyristor," *IEEE Trans. Electron Devices*, **ED-22**, No. 2, pp. 65–68 (1975).

19. I. V. Grekhov et al., "Investigation of the Propagation of the Turned-On State Along a *p-n-p-n* Structure," *Soviet Phys.—Semicond.*, **4**, No. 11, pp. 844–849 (1971).

20. W. H. Dodson and R. L. Longini, "Skip Turn-On of Thyristors," *IEEE Trans. Electron Devices*, **ED-13**, No. 7, pp. 598–604 (1966).

21. H. F. Storm and J. G. St. Clair, "An Involute Gate-Emitter Configuration for Thyristors," *IEEE Trans. Electron Devices*, **ED-21**, No. 8, pp. 520–522 (1974).

22. F. E. Gentry and J. Moyson, "The Amplifying Gate Thyristor," paper no. 19.1, IEEE Meeting of the Professional Group on Electron Devices, Washington, D.C., 1968.

23. P. Voss, "A Thyristor Protected Against *di/dt* Failure at Breakover Turn-On," *Solid State Electron.*, **17**, No. 7, pp. 655–661 (1974).

24. P. S. Raderecht, "A Review of the Shorted Emitter Principle," *Int. J. Electron.*, **31**, 1st Ser., No. 6, pp. 541–564 (1971).

25. A. Munoz-Yague and P. Leturcq, "Optimum Design of Thyristor Gate-Emitter Geometry," *IEEE Trans. Electron Devices*, **ED-23**, No. 8, pp. 917–924 (1976).

26. T. C. New and D. E. Cooper, "Turn-On Characteristics of Beam Fired Thyristors," *IEEE Conference Record*, Industry Applications Society, pp. 259–265, 1973.

27. A. A. Bergh and P. J. Dean, "Light Emitting Diodes," *Proc. IEEE*, **60**, No. 2, pp. 156–223 (1972).

28. R. D. Maurer, "Glass Fibers for Optical Communication," *Proc. IEEE*, **61**, No. 4, pp. 452–462 (1973).

29. C. A. Burns and B. I. Miller, "Small Area DH Al GaAs Electroluminescent Diode Sources for Optical Fiber Transmission Lines," *Opt. Commun.*, **4**, No. 3, pp. 307–309 (1971).

30. V. A. K. Temple and A. P. Ferro, "High-Power Dual Amplifying Gate Light Triggered Thyristor," *IEEE Trans. Electron Devices*, ED-23, No. 8, pp. 893–898 (1976).

31. D. Silber et al., "Progress in Light Activated Power Thyristors," *IEEE Trans. Electron Devices*, **ED-23**, No. 8, pp. 899–904 (1976).

32. T. S. Sundresh, "Reverse Transient in *p-n-p-n* Tridoes," *IEEE Trans. Electron Devices*, **ED-14**, No. 7, pp. 400–402 (1967).

33. I. Somos, "Switching Characteristics of Silicon Power Controlled Rectifiers," *IEEE Trans. Commun. Electron.*, **83**, No. 75, pp. 861–871 (1964).

34. J. B. Brewster and E. S. Schlegel, "Forward Recovery in Fast Switching Thyristors," *IEEE Conference Record*, Industry Applications Society Meeting, pp. 663–672, 1974.

35. E. Schlegel, "Gate-Assisted Turnoff Thyristors," *IEEE Trans. Electron Devices*, **ED-23**, No. 8, pp. 888–892 (1976).

36. J. Shimizu et al., "High-Voltage High-Power Gate-Assisted Turn-Off Thyristor for High Frequency Use," *IEEE Trans. Electron Devices*, **ED-23**, No. 8, pp. 883–887 (1976).

37. T. Matsuzawa and Y. Usunaga, "Some Electrical Characteristics of a Reverse Conducting Thyristor," *IEEE Trans. Electron Devices*, **ED-17**, No. 9, 816 (1970).

38. H. Oka and H. Gamo, "Electrical Characteristics of High Voltage, High Power, Fast Switching Reverse Conducting Thyristor and Its Applications for Chopper Use," *Proc. IEEE Conference on Power Converters*, pp. 1.5.1–1.5.12, 1973.

39. B. Asalit and G. H. Studtmann, "Description of a Technique for the Reduction of Thyristor Turn-Off Time," *IEEE Trans. Electron Devices*, **ED-21**, No. 7, pp. 416–420 (1974).

40. E. D. Wolley, "Gate Turn-off in *p-n-p-n* Devices," *IEEE Trans. Electron Devices*, **ED-13**, No. 7, pp. 590–597 (1966).

41. Y. C. Kao and J. C. Brewster, "A Description of the Turn-Off Performance of Gate Controlled Switches," *IEEE Conference Record*, Industry Applications Society Meeting, pp. 689–693, 1974.

42. E. D. Wolley et al., "Characteristics of a 200 Amp Gate Turn-Off Thyristor," *IEEE Conference Record*, Industry Applications Society Meeting, pp. 251–255, 1973.

43. F. E. Gentry et al., "Bidirectional Triode *P-N-P-N* Switches," *Proc. IEEE*, **53**, No. 4, pp. 355–369 (1965).

44. J. F. Essom, "Bidirectional Triode Thyristor Applied Voltage Rate Effect Following Conduction," *Proc. IEEE*, **55**, No. 8, pp. 1312–1317 (1967).

45. S. K. Ghandhi, *The Theory and Practice of Microelectronics*, John Wiley & Sons, New York, 1968.

46. J. S. T. Huang, "Lateral *p-n-p-n* Device," *Solid State Electron.*, **11**, No. 8, pp. 779–785 (1968).

47. H. F. Storm and A. P. Ferro, "A Bilateral Silicon Switch," *IEEE Trans. Electron Devices*, **ED-14**, No. 6, pp. 330–333 (1967).

48. K. Hubner et al., "Uniform Turn-On in Four Layer Diodes," *IRE Trans. Electron Devices*, **ED-8**, No. 6, pp. 461–464 (1961).

49. P. Voss, "Observation of the Initial Phases of Thyristor Turn On," *Solid State Electron.*, **17**, No. 8, pp. 879–880 (1974).

5.10 PROBLEMS

1. Determine the reverse blocking characteristics for a thyristor, where $W_{N1} = 100$ μm, $L_{pN1} = 100$ μm, and $\gamma = 1$. What is the optimum starting resistivity for this device?

2. Calculate the reverse blocking characteristic for the device of Problem 1, assuming that $\gamma = 0.7$ and is constant.

3. Calculate the forward blocking characteristic of the thyristor of Problem 1, assuming that the low level current gain of the *n-p-n* transistor is 0.5 and is constant. Repeat for values of $\alpha_{npn} = 0.1$ and 0.7, and plot on the same graph.

4. Consider a thyristor design of Fig. 5.5, where $N_{N1} = 7 \times 10^{13}/\text{cm}^3$. This thyristor is operated with $V_{AK} = 500$ V. Compute the undepleted width of the $N1$ region as a function of gate current. Calculate the forward current as a function of gate current, for any fixed value of leakage current. Assume that $\gamma_{npn} = 1$ and that forward breakover is due only to changes in the base transport factor of the *p-n-p* transistor.

5. Make a detailed geometrical drawing of an involute gate-cathode structure, having four gate fingers. Show clearly the gate-cathode separation. Assume a 50 mm diameter slice, with 1 mm wide gate fingers, and 0.1 mm gate-cathode separation.

6. A 20 mm diameter thyristor is to be designed to have a dv/dt rating of 200 v/μsec. Shunts of 0.5 mm diameter are to be used. Determine the separation between shunts and the total number of shunts that are required. Assume that the depletion layer capacitance of the J_2 junction is 300 pF, $W_{P2} = 25$ μm, and $\rho_{P2} = 5$ ohm-cm.

7. What is the dv/dt rating of the thyristor of Problem 6, if the separation between shunts is doubled? For the case when it is halved?

8. Determine the dv/dt for the device of Problem 6 for $\rho_{P2} = 1$ and 10 ohm-cm, respectively.

Fabrication Technology

CONTENTS

POWER SEMICONDUCTOR DEVICES are silicon based. Thus techniques for device fabrication are very similar to those used for the manufacture of integrated circuits. For example, power devices use slices of silicon, appropriately cut and lapped, as the starting material. Device processes such as epitaxy, diffusion, and alloying are employed, sometimes in conjunction with oxidation and photomasking, to form the appropriate junctions in these slices. After pelletizing, leads are attached and the device is die bonded to a header, then suitably encapsulated.

Notwithstanding the similarities just mentioned, the power device is sufficiently different from the microcircuit that the emphasis on various fabrication steps is often quite different. Thus we have the following conditions.

1. Power devices are quite large and frequently utilize an entire slice for a single device. This imposes stringent requirements on the quality and uniformity of the starting material. In addition, device bonding to the package must be done with careful attention to stress relief and to the minimization of fatigue from thermal cycling.

2. Power devices operate at high voltages. This necessitates the use of special pellet treatments such as surface contouring, and surface coating with glasses and silicones.

3. Carrier lifetime in power devices is usually required to be large (10–30 μsec), to minimize forward voltage drop. In addition, deep diffusions, necessitating long diffusion times, are generally used. These combined requirements place great importance on high temperature processing steps, which must be conducted under extremely clean environmental conditions. In addition, gettering techniques are very important in many power device processes.

At the present time, devices are routinely available in 52 and 76 mm diameters (2 and 3 in., nominal), and 100 mm diameter (4 in. nominal) devices have already appeared in the marketplace. Surge current ratings as high as 24,000 A in a single device are already achievable in these 100 mm devices, allowing them to be operated without ancillary protective components such as fuses or circuit breakers. The ability to sustain reverse voltages has also improved; 3000 V devices are routinely available, and it appears that this limit can be further extended to 5000–10,000 V.

The technology of conventional monolithic integrated circuits, which serves as a convenient starting point for this chapter, has been described in detail elsewhere [1]. Our emphasis is placed on the aspects of power device fabrication technology that differ significantly from these conventional processes.

6.1 STARTING MATERIALS

Integrated circuit technology requires relatively low starting resistivity material (3–10 ohm-cm), which is primarily used as a substrate on which epitaxial layers are grown for functional elements such as transistors and resistors. As a result, the properties of the silicon wafer are of secondary importance, the burden of device performance being placed on the epitaxial layer. Furthermore, integrated circuits are generally no larger than 0.5×0.5 cm, and large slices are used only because of the cost improvements that can be achieved by processing many circuits on a single wafer. In contrast, power semiconductors use the bulk material and are large—in many instances a single device occupies an entire wafer. Local inhomogeneities, which may result in the rejection of some individual dice on an integrated circuit wafer, often cause the rejection of an entire silicon slice.

6.1.1 Czochralski Silicon

At present reasonably defect-free silicon can be most economically grown in the $\langle 111 \rangle$ crystallographic orientation. The starting material, chemically purified silicon, can be melted and grown into single crystal form by the Czochralski (CZ) process, shown schematically in Fig. 6.1. Here the silicon is contained in a quartz or graphite crucible and kept in a molten condition by radio frequency (r.f.) or by resistive heating. A seed crystal is suspended over the crucible in a chuck. For growth, the seed is inserted into the melt until its end is molten, then withdrawn at a rate of about 10 μm/sec, resulting in a single crystal that grows by progressive freezing at the liquid-solid interface. Provisions are usually made to rotate both the crucible and the crystal during the pulling operation.

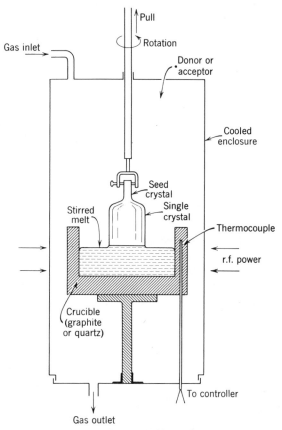

Fig. 6.1   Czochralski apparatus.

The entire assembly is enclosed within a chamber, which sometimes has water-cooled walls; then it is flushed with an inert gas such as helium or argon. A feedback control system maintains the temperature of the melt to within $\pm 0.5°C$, since thermal conditions are always changing during this operation.

The CZ technique can be readily used to grow single crystal silicon with resistivity as high as 25–50 ohm-cm. By suitable programming of spin and pull rates, this resistivity range can be achieved over 80% of the ingot length. Crystal diameters as large as 100 mm are routinely grown with fewer than 500 dislocations per square centimeter, in ingots that weigh more than 10 kg. Almost all silicon used in integrated circuit manufacture is grown by this technique. However CZ-grown material has a number of characteristics that create problems in power device manufacture, as follows.

1. Many power devices require starting resistivities well in excess of 50 ohm-cm. Even after repeated melting and regrowth, CZ crystals can only be made with resistivities as high as 100–200 ohm-cm, because of contamination from the walls of the crucible in which the melt is held. Consequently the use of this material is restricted to devices with breakdown voltages below 750 V.
2. The solid-liquid interface is generally not maintained flat during CZ crystal growth, but tends to take on an Ω-shape. This results in radial forces and preferential alignment of both mechanical and chemical defects (i.e., dopants). Consequently both radial resistivity and radial defect concentration variations are present in this material. The breakdown voltage rating of devices made on this material is thus lower than that given by the nominal resistivity value, in the absence of these variations. Radial resistivity variations in excess of $\pm 25\%$ are typically obtained in high resistivity silicon grown by the CZ process.
3. Carbon is often a dissolved impurity in the starting material. This element tends to remain in the silicon during crystal growth and is present in the form of silicon carbide precipitates and clusters, which can be as large as 0.1–5 μm. Although generally insulating in nature, these "macroprecipitates" can cause curvature of potential lines during device operation, as well as localized enhancement of the electric field in their immediate vicinity. Thus they give rise to the formation of soft spots and premature breakdown [2].
4. Commonly grown in silica crucibles, CZ crystals contain oxygen, which is dissolved to its solid solubility limit (10^{18} atoms/cm^3) by the corrosive action of the molten silicon against the crucible wall. Most of this oxygen is present as electronically inactive SiO_2, which segregates

preferentially at sites of vacancy clusters [3], forming macroprecipitates about 1.0 μm in diameter and about 1.0 μm apart. Once again, this can give rise to the formation of soft spots in the silicon bulk.

Even if oxygen-free silicon is used as the starting material, it is important to prevent the entry of oxygen during high temperature device processing. Thus oxide-masking processes, which are universally used in integrated circuit technology, are not employed in the manufacture of high voltage power devices. Alternate masking techniques, such as the use of silicon nitride, are to be preferred in these cases.

Single crystal silicon contains many impurities in addition to carbon and oxygen. These include a number of heavy metals (iron, manganese, cobalt, copper, nickel, etc.), some of which come from the crucible walls. Their solid solubility is quite small, typically less than 10^{15} atoms/cm^3. Furthermore, usually less than 0.1% of these impurities is found to be electronically active in silicon. Most, however, are in the form of microprecipitates and silicides, usually smaller than 0.1 μm. Processing conditions that introduce considerable strain can cause them to move and cluster, ultimately coalescing into macroprecipitates. As much as possible, such processes should be avoided during power device manufacture.

6.1.2 Float Zone Silicon

An alternate method for single crystal silicon growth is the float zone (FZ) process, presented schematically in Fig. 6.2. The FZ process uses as the starting material a rod of cast silicon, which is usually maintained in a vertical position and rotated during the operation. A small zone (typically 1.5 cm long) of the crystal is kept molten by means of r.f., and the r.f. coil moved to permit this floating zone to traverse the length of the bar. A seed crystal is provided at the starting point where the molten zone is initiated and is arranged so that its end is just molten. As the zone traverses the bar, single crystal silicon freezes at its retreating end and grows as an extension of the seed crystal. Often more than one molten zone is used during this process.

The FZ process is ideally suited to crystal refinement because of the ease with which multiple zone passes can be made. Figure 6.3 illustrates the successive compositions of a crystal with a rather high value of effective distribution coefficient, $k_e = 0.1$, after a number of such passes [4]. It is seen that even here* the degree of crystal refinement obtained by this method is

*Typical values of k_e for impurities in silicon are usually well below 0.1. Thus the degree of crystal refinement is higher for most impurities than is shown here.

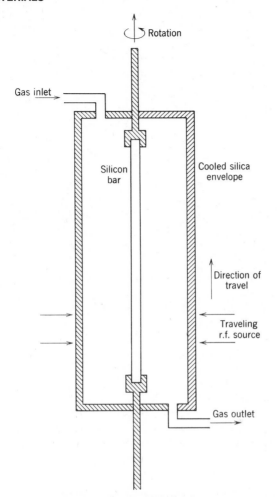

Fig. 6.2 Float zone apparatus.

very high. Typically the FZ process readily provides material in the 3000 ohm-cm range (*p*-type) and in the 1000 ohm-cm range (*n*-type). Resistivities as high as 12,000 ohm-cm (*p*-type) and 4000 ohm-cm (*n*-type) are achievable by this method.

Single crystal silicon is unsupported during growth by the FZ process. As a result, many volatile impurities can be removed by crystal growth in a reduced pressure atmosphere. In addition, impurity contamination from the crucible walls is eliminated by this process. Table 6.1 lists the typical impurity contents of two crystals, of identical resistivity (15 ohm-cm),

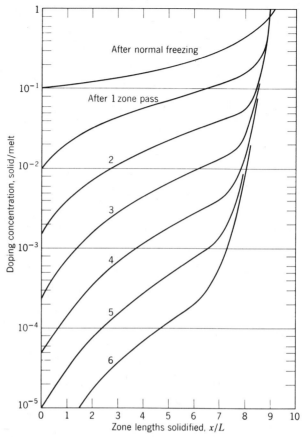

Fig. 6.3 Doping profile with multiple passes. From *Transistor Technology*, Vol. 1, by H. E. Bridges et al. Copyright 1958. Used with permission of Van Nostrand Reinhold Company.

Table 6.1 Impurity Content of Typical Silicon Crystals

Impurity	Float Zone	Czochralski
C	2.2×10^{16}	1.5×10^{17}
O	3×10^{16}	4×10^{17}
F	1×10^{16}	2×10^{16}
Na	1.5×10^{15}	10^{15}
Ca	2.71×10^{15}	2×10^{16}
Cl	10^{15}	10^{15}
Fe	$< 10^{15}$	10^{16}
Cu	0.85×10^{16}	1.9×10^{16}
K, Co, Mn, Ni	$< 10^{15}$	$< 10^{15}$

grown by the Czochralski and float zone processes [5]. Of particular interest are the greatly reduced levels of oxygen and carbon resulting from the FZ process. As mentioned earlier, both these impurites tend to form macroprecipitates in silicon, resulting in premature breakdown in high voltage devices.

In view of these advantages, FZ silicon is being increasingly used by the power semiconductor industry, its slightly higher cost than CZ material notwithstanding. It is the exclusive choice for devices that must sustain reverse potentials in excess of 750 V.

6.1.3 Neutron Doping

A major problem with float zone material is the presence of radial microresistivity variations. These take the form of large, abrupt resistivity changes, which occur over intervals 10–100 μm wide and can be detected only by high resolution spreading-resistance probe techniques.

Microresistivity variations are generally considered to be caused [6] by a misalignment of the rotational axis of the crystal and the thermal axis of symmetry of the crystal growth apparatus. Although commonly encountered in FZ material, where they show up in the form of growth striations and swirls, they have also been noted in CZ crystals, where they are usually obscured because of its higher background doping level. They can be reduced, but not eliminated, by careful mechanical design of the FZ apparatus.

Microresistivity variations can be virtually eliminated by a new crystal preparation technique known as *neutron doping* [7]. This technique is based on the fact that bombardment of silicon with thermal neutrons gives rise to the formation of finite amounts of phosphorus, which is an *n*-type dopant. Specifically, the ^{30}Si isotope form occurs as a component of native silicon, with a concentration of about 3%. Neutron bombardment causes this ^{30}Si to change to ^{31}Si, liberating gamma rays; simultaneously, the ^{31}Si undergoes a transition to phosphorus (^{31}P) with the liberation of beta rays, with a half-life of 2.62 hr.

A secondary process results in the conversion of ^{31}P to ^{32}P, which transmutes to ^{32}S. This reaction has a half-life of 14.3 days and is present in significant quantities only for heavy doping. Typically, neutron doping down to 10 ohm-cm can be accomplished with a cool-down period of 3–4 days, to be completely safe by international standards.

Since the penetration range of neutrons in silicon is roughly 90–100 cm, doping is very uniform throughout the slice. Thus starting with slices of FZ silicon having an average resistivity well in excess of what is required, it is possible to obtain uniformly doped slices, free from both macro- and microresistivity variations. Figure 6.4 shows spreading resistance (R_S) data for a typical 100 ohm-cm (nominal) slice of grown silicon, as well as for a slice prepared by the neutron doping technique. Typically, radial resistivity variations can be reduced to about ± 1% by this technique.

Annealing for as little as 2 min at 750°C has been found [8] to be sufficient to stabilize the neutron-doped silicon against resistivity changes. Radiation damage of crystals does not present a problem, since normal crystal processing during device fabrication is entirely sufficient to anneal out all defects created in this manner.

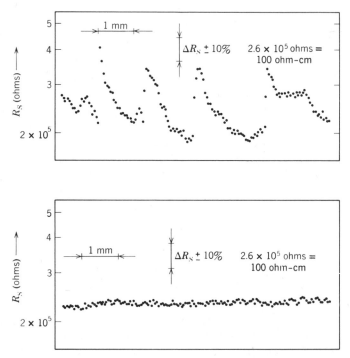

Fig. 6.4 Microresistivity variations. (a) Conventional growth. (b) Neutron doped. From Herrmann and Herzer [4]. Reprinted with permission of the publisher, The Electrochemical Society, Inc.

By its very nature, radiation doping is restricted to n-type silicon. This is not a serious limitation, however, since nearly all power devices are made with n-type starting material. A number of major suppliers of silicon crystal now have installed radiation doping facilities, and this material should become the universal choice for high power semiconductor devices.

6.2   JUNCTION FORMATION

Both p- and n-type shallow impurities are used in power semiconductors for the formation of junctions. For silicon, the n-type impurities are phosphorus, arsenic, and antimony. Of these, microcircuit technology has favored phosphorus for conventional doping applications, and arsenic or antimony when a particularly slow moving impurity is required. The p-type impurities are boron, aluminum, and gallium. Boron is the only dopant used in microcircuit technology, since neither aluminum nor gallium can be photomasked by silicon dioxide.

The basic techniques for junction formation are diffusion, alloying, and epitaxy. Each has its own characteristics and is used on different occasions in the fabrication of power semiconductor devices. Most of these techniques can also be used for the formation of ohmic contacts. However this topic is dealt with separately in Section 6.4 because of the slightly different requirements to be met.

6.2.1   Diffusion

Many regions of power semiconductor devices are fabricated by diffusing into the silicon slice. Diffusion depths are ordinarily much larger than those used with conventional devices, often exceeding 25 μm. Thus high diffusion temperatures and long diffusion times are commonly encountered, especially in high voltage structures.

An important requirement for dopant introduction in power semiconductors is that diffusion-induced defects be minimized. Thus the diffusion process should not create a large stress in the semiconductor; alternately, techniques must be used for minimizing this stress. One such technique is to dope semiconductor slices symmetrically on both sides, subsequently lapping off one side, as in Fig. 6.5, which illustrates the sequence of steps used in forming a double diffused,[*] high voltage n^+-p-ν-n^+ power transistor. Here the starting material is a slice of lightly doped ν-type silicon that is symmetrically doped with a p-type impurity (Fig. 6.5a). Next, one side of the slice is lapped off, resulting in the structure of Fig. 6.5b. Finally, the slice is symmetrically doped with an n^+-impurity to result in the n^+-p-ν-n^+ structure of Fig. 6.5c. Boron and phosphorus are commonly used as diffusants in devices of this type.

The diffusion of an impurity will introduce a stress in the crystal lattice if its tetrahedral radius is different from that of the host silicon. This difference is characterized by a *misfit factor* ε, defined by $r = r_0 (1 \pm \varepsilon)$,

---

[*]This is sometimes referred to in the literature as a "triple diffused" structure because it results in three diffused regions.

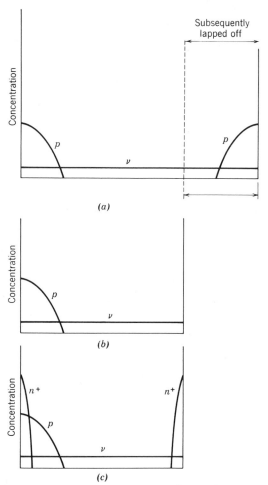

Fig. 6.5   Fabrication steps for an n^+-p-ν-n^+ transistor.

where r and r_0 are the tetrahedral radii of the impurity and of silicon, respectively ($r_0 = 1.18\text{Å}$). Lattice contraction or dilation occurs if the impurity atom is smaller or larger than the silicon atom. The magnitude of the stress created in this manner depends on the solid solubility of the impurity at the diffusion temperature, as well as on its misfit factor. The elastic limit of silicon can even be exceeded in some cases, giving rise to the formation of dislocations. These diffusion-induced dislocations can, in turn, affect the fabrication yield, since they form sites for the condensation of heavy metallic impurities and also for vacancy generation.

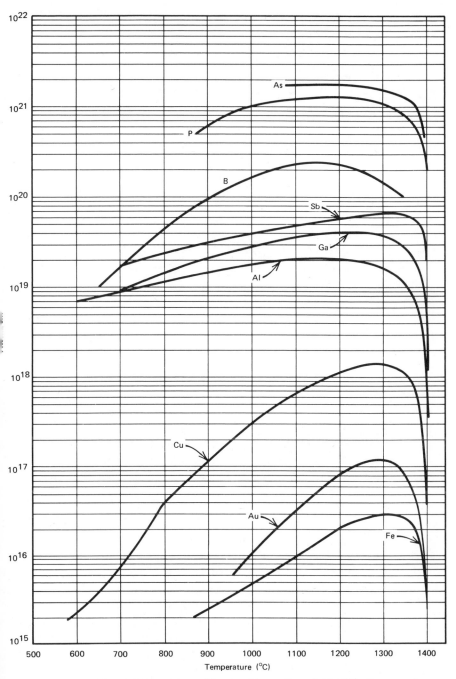

Fig. 6.6 The solid solubility of impurities in silicon. Copyright 1960, the American Telephone and Telegraph Company. Reprinted with permission from Trumbore [9].

Table 6.2 Properties of Shallow Impurities in Silicon

Property	p-type Al	B	Ga	n-type As	P	Sb
Ionization energy: distance from valence band (eV)	0.057	0.045	0.065	—	—	—
Ionization energy: distance from conduction band (eV)	—	—	—	0.049	0.045	0.039
Tetrahedral radius, r (Å)	1.29	0.98	1.27	1.18	1.07	1.36
Misfit factor, ε	.068	.254	.068	0	.068	.153
Stress $\times 10^{-9}$ dynes/cm^2 for a 1200°C diffusion (+ lattice contraction − lattice dilation)	−0.2	+7	−0.4	0	+10.4	−3

Figure 6.6 shows the solid solubility [9, 10] of a number of impurities in silicon. Table 6.2 lists the common shallow impurities, their energy levels, tetrahedral radii, misfit factors, and the stress created by their diffusion into silicon at 1200°C. Note that the elastic limit of silicon, 1×10^9 dynes/cm^2 at this temperature, is exceeded in a number of cases illustrated here [11].

Techniques for stress-free diffusion are under active study, and they involve simultaneous diffusion of the dopant with an inert inpurity such as tin [12] for stress control. Other approaches emphasize the use of arsenic as a doping impurity, since it is a perfect match to the silicon lattice. The primary effort in this area has been aimed at the development of microwave transistors, for which extremely tight control of shallow junction depths is mandatory. Arsenic diffuses relatively slowly into silicon, as compared to phosphorus. This is a distinct disadvantage in power semiconductors, where junction depths are usually large (>5–10 μm, vs. 0.5–1 μm in microwave transistors).

Power devices often require deep, contamination-free diffusions to obtain junctions with high reverse breakdown voltage and long lifetime. Long diffusion times at high temperatures are commonly encountered. For example, 48 hr diffusions at 1200°C are typical during high voltage thyristor manufacture. A simultaneous requirement is that these diffusions have a surface concentration sufficiently low to permit them to be followed by a diffusion of the opposite conductivity type. This surface concentration is, in turn, dictated by the solid solubility of the dopant at the diffusion temperature, and by its vapor pressure.

Both gallium and aluminum are suitable p-type impurities for this type of diffusion. Both are relatively fast diffusers (as compared to boron, which is the only other p-type dopant). In addition, their use in sealed tube diffusion systems results in relatively low surface concentrations by a simple high temperature, one-step process.* Of these, gallium is less reactive than aluminum and is more commonly used.

An interesting feature of gallium is its ready transport through layers of silicon dioxide. Thus it cannot be selectively diffused by photomasking of oxide layers. By the same token, the use of gallium results in very parallel plane, uniform diffusions into both sides of the silicon slice, since it is unaffected by patches of native SiO_2 that are normally always present on the silicon surface.

Aluminum has a somewhat higher diffusion constant than gallium. Its introduction in combination with gallium results in the formation of deep p-regions with sufficiently high doping concentration, but with greatly reduced processing time.

6.2.1.1 Sealed Tube Systems

The first diffusion step in the fabrication of high voltage thyristors consists of making deep p-type diffusions into both sides of a starting n-type silicon slice. A sealed tube system is most suited for this diffusion, to avoid contamination. Here a large number of n-type slices, suitably lapped and cleaned, are loaded into a boat and placed in a quartz tube, together with elemental gallium as the dopant source. The tube is evacuated, back-filled with low pressure argon, sealed off, and placed in the diffusion furnace for the appropriate time and at a suitable temperature. Upon completion of the diffusion, the tube is returned to room temperature and cut open. Slices diffused in this way have a p-type region formed on the entire silicon surface. Thus the rim must be eventually cut off to provide a p-n-p structure. This is usually done by chemical lapping and grinding methods, when the appropriate bevel angles are incorporated into the device.

The vapor pressure of elemental gallium is extremely low at diffusion temperatures (typically 2×10^{-1} mm Hg at 1200°C). Thus the gallium surface concentration in the silicon slice is considerably lower than its solid

---

*Two-step diffusions are routinely used in microcircuit fabrication to circumvent this problem. However they are considerably more complex, as well as more difficult to control for deep-junction formation.

solubility. A surface concentration of about 5×10^{18} atoms/cm^3 is ordinarily obtained, with typical diffusion times and temperatures of 48 hr and 1200°C for a high voltage device.

The simultaneous diffusion of gallium and aluminum is carried out in the same manner, except that elemental aluminum is used in addition to the gallium source. Aluminum diffusions are difficult to control because this element is highly reactive. Thus special care must be taken [13] to ensure the complete absence of moisture or oxygen in the diffusion tube before sealing. This is necessary because the aluminum source must be maintained in its elemental form for its vapor phase transport to the silicon surface. High temperature baking of the quartz tube, as well as the use of alumina liners, both reduce the chances for source oxidation and improve the surface concentration that can be achieved by this type of diffusion. Even under optimum conditions, however, the vapor pressure of aluminum is well below that of gallium (by a factor of 20 at 1200°C), and the achievable surface concentration is lower by a factor of 10–20. Typical values of surface concentration for aluminum alone range from 3×10^{16} to 8×10^{17} per cubic centimeter, depending on the diffusion conditions. Its use in conjunction with gallium results in increasing this value to 5×10^{18}/cm^3, which is ordinarily obtained with gallium alone.

6.2.1.2 Open Tube Systems

Open tube diffusion systems are more convenient to operate than sealed systems, but they are subject to impurity contamination and require high levels of cleanliness for satisfactory operation. Their use is consequently restricted to the fabrication of relatively shallow junctions (under 25 μm), or to situations in which long lifetimes are not essential, as in transistor manufacture.

The open tube system (Fig. 6.7) consists of a dopant source and carrier gas handling apparatus, feeding a resistance heated tube of high purity quartz. Temperature control is maintained to ± 1°C over a flat zone that is 24 in. long in many production systems.

A large variety of dopant sources have been used in microcircuit manufacture. For power devices, however, attention has been primarily concentrated on halogenic compounds of the dopants. Phosphorus oxychloride (POCl$_3$) is commonly used for n-type diffusions. This liquid source is kept in a bubbler maintained at 0–40°C, and its vapors are transported to the silicon slices in the presence of oxygen. The POCl$_3$ is converted upstream into phosphorus pentoxide and chlorine. Phosphorus pentoxide forms a glass on the silicon slice, which is reduced to phosphorus by the silicon, and enters the crystal lattice. The chlorine leaves by

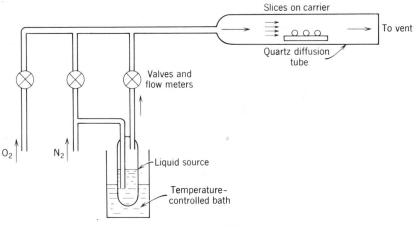

Fig. 6.7   An open-tube diffusion system.

a suitably vented exit port. Reactions associated with this process are as follows:

Upstream

$$4POCl_3 + 3O_2 \rightarrow 2P_2O_5 + 6Cl_2$$

At the silicon slice

$$2P_2O_5 + 5Si \rightleftharpoons 5SiO_2 + 4P$$

Boron tribromide (BBr_3) is commonly used for p-type diffusions. This liquid source is typically held at 0°C and its vapors transported to the silicon in an oxygen-rich carrier gas ambient. Boron trioxide (B_2O_3) is formed upstream and deposits as a glassy layer on the silicon surface. This layer acts as a source of elemental boron for diffusion into the silicon lattice. Bromine, resulting from the upstream reaction, leaves by the exit port. Reactions describing this process are as follows:

Upstream

$$4BBr_3 + 3O_2 \rightarrow 2B_2O_3 + 6Br_2$$

At the silicon slice

$$2B_2O_3 + 3Si \rightleftharpoons 3SiO_2 + 4B$$

An interesting technique for making diffusions with surface concentrations below the solid solubility is to use [14] dilute, doped oxide sources. These materials, consisting of oxides of the dopant and silicon dioxide, suspended in a volatile liquid binder, are commercially available. In practice, a few drops of this dopant is placed on the surface of a silicon slice, which is rotated at high speed. Evaporation of the binder results in the *in situ* formation of the diffusant source. Spin-on dopants of this type are very promising for small, low voltage thyristors, where many devices are fabricated on a single slice by photomasking processes and the effects of oxygen in the silicon slice are not important. Boron can be used as a dopant in this manner, to form masked *p*-type diffusions with low surface concentration.

6.2.1.3 Doping Profiles

The doping profile that results from the processes just outlined is of the complementary error function type for a single dopant. Taking the origin at the surface, this profile is described by

$$N(x,t) = N(0,t)\text{erfc}\left(\frac{x}{2\sqrt{Dt}}\right) - N_B \qquad (6.1)$$

where $N(x,t)$ = impurity concentration in silicon (atoms/cm^3) at a depth
  of x cm from silicon surface, at time t
$N(0,t)$ = impurity concentration at surface (atoms/cm^3) at time t
D = diffusion constant for particular operating temperature
  (cm^2/sec), shown in Fig. 6.8 for shallow impurities in
  silicon
N_B = background concentration of opposite conductivity type
  (atoms/cm^3)
t = time (sec)

The complementary error function term in this equation is plotted in Fig. 6.9 in the normalized form, $y = \text{erfc } z$. Note that this curve can be approximated by a simple exponential at doping levels below 1% of the surface concentration. A convenient equation for this profile has been given in (2.54) as

$$N(x) = N_B\left(e^{-x/\lambda} - 1\right) \qquad (6.2)$$

where λ is the space constant and the origin is taken at the metallurgical junction.

The doping profile for simultaneous diffusion from gallium and aluminum can be considered to be the superposition of two independent complementary error function diffusions into the background concentration. The surface concentration of the aluminum is typically much lower

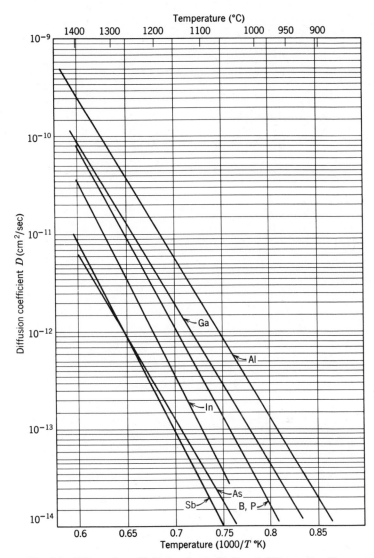

Fig. 6.8 **Diffused coefficients of substitutional diffusers in silicon.**

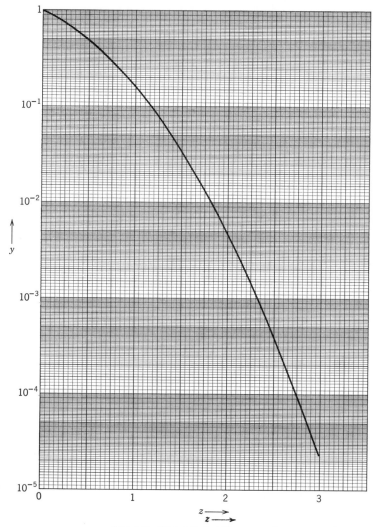

Fig. 6.9 The erfc function $y = \text{erfc } z$.

than that of gallium, by a factor of 20 or more; thus its primary effect is in reducing the diffusion time.

6.2.2 Alloying

The alloyed junction is formed at relatively low temperatures and is one of the later steps in the fabrication of a multijunction device. Alloying is usually done to a metal such as gold, with which silicon forms a relatively low melting point eutectic. With reference to the phase diagram of the gold-silicon system (Fig. 6.10), pure gold and pure silicon melt at 1063 and 1412°C, respectively. However the eutectic composition, 94% gold and 6% silicon by weight, melts at 370°C. Under conditions of intimate contact, therefore, gold dissolves into the silicon to form an alloyed region with a composition close to that of the eutectic. Intimate contact of this type is best accomplished by plating the silicon surface with gold. Alternately, it is possible to place a thin gold disc, known as a preform, in contact with the silicon, to obtain alloying by means of pressure, heat and vibration.

The use of a preform of the *eutectic composition* lowers the temperature at which this process can be accomplished, and it is a preferred approach. In practice, gold-germanium eutectic preforms (88% gold, 12% germanium by weight) are often used because of their better wetting characteristics and slightly lower eutectic melting temperatures (356°C, as seen in Fig. 6.11).

Actual junctions are formed only if the alloyed material is of the opposite conductivity type to the silicon. This is accomplished by predoping the preform with about 0.1% antimony or gallium (*n*- or *p*-type) as

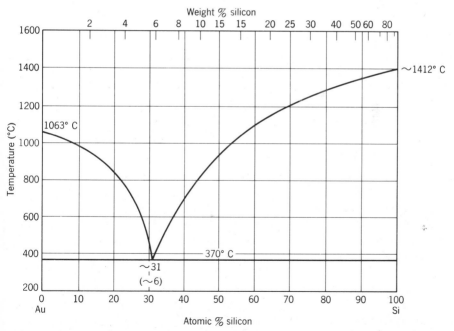

Fig. 6.10   The gold-silicon system. From *Constitution of Binary Alloys*, by M. Hansen and A. Anderko [15]. Copyright 1958. Used with permission of the McGraw-Hill Book Company.

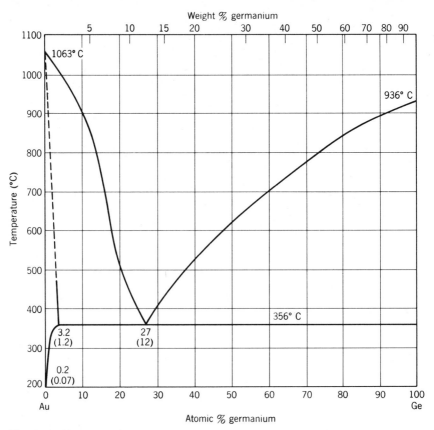

Fig. 6.11 The gold-germanium system. From _Constitution of Binary Alloys_, by M. Hansen and A. Anderko [15]. Copyright 1958. Used with permission of the McGraw-Hill Book Company.

required. Alloyed regions of this type are heavily saturated with the dopant and are thus degenerate n^+- or p^+-type, respectively.

The cathode region of many p-n-p-n thyristors is made by alloying. Here a disk-shaped preform is placed on the surface of the freshly cleaned p-type silicon surface and held in place by means of a suitable jig. The assembly is placed on the moving belt of a furnace through which it passes in a slightly reducing (20% hydrogen, 80% nitrogen) atmosphere. An antimony-doped preform is used in this step, to make the cathode region highly n-type. The process is a continuous one, with loading at one end and unloading at the other; a typical pass through a furnace of this type takes only a few minutes. The ohmic contact to the gate region is commonly made at the same time, during this pass.

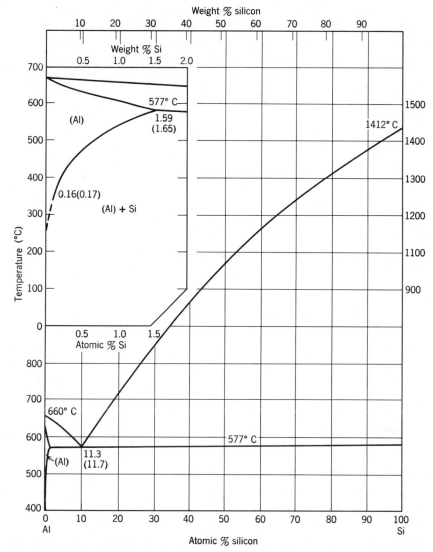

Fig. 6.12 **The aluminum-silicon system. From *Constitution of Binary Alloys*, by M. Hansen and A. Anderko [15]. Copyright 1958. Used with permission of the McGraw-Hill Book Company.**

The aluminum-silicon system is often used for the formation of alloyed junctions. As Fig. 6.12 indicates, it exhibits a eutectic temperature of 577°C, whereas aluminum melts at 660°C. In this case, the alloyed region is saturated with aluminum and is degenerately p^+-type. Thus its use in junction formation is suitable for p^+-n structures or in p^+-ν-n^+ diodes.

6.2.3 Epitaxy [16]

Epitaxy is the process of setting an amount of material on top of a crystalline substrate while still preserving the overall single crystal structure. As applied to the fabrication of power semiconductors, it describes the process of growing a relatively thin (5–50 μm) single crystal layer of suitably doped silicon on a single crystal silicon substrate. Junctions can be formed if the grown layer is of opposite conductivity type.

Epitaxial growth provides an alternative to diffusion as a process for fabricating appropriately doped semiconductor regions. Unlike diffusion, each new layer is quite independent of the last and is not the result of a counterdoping process. Consequently the total impurity concentration of each layer can be maintained at a reasonable level, and there is no limit to the number of successive layers that can be grown.

Fundamental to this process is a means for transporting atoms of the material to be grown to the substrate on which epitaxy is desired. On arriving on this substrate, the atoms move around until they find a region to which they can attach. This movement is enhanced because the substrate is maintained at an elevated temperature. Attachment of the arriving atoms occurs preferentially at a nucleation center. Crystal growth is initiated at a number of such centers and spreads laterally until a layer is completed. In principle, fresh nucleation centers are formed on this layer and the process repeats. In this manner, the epitaxial film is built up as a series of atomic planes. The growing layer follows the substrate plane sequence, since this results in an energy minimum for the system as a whole.

In practice it is impossible to confine the epitaxial growth to an atomic layer at a time. In fact, better crystal quality is obtained if the epitaxial layer is grown on a slightly misoriented crystallographic plane. This encourages the nucleation process to occur at the step corners formed by the (111) plane surfaces and allows a higher deposition rate without the development of surface defects. Typically, a 2°–3° misorientation toward the (110) plane allows the epitaxial growth rate to be doubled with the same crystal perfection.

It is common practice to dope the layer during the growth process. As a result, the silicon atoms are accompanied by impurity atoms that are usually of different ionic radii. This may lead to disorder in the crystal structure of the epitaxial layer.

Silicon atoms are delivered to the substrate by an indirect process involving the hydrogen reduction of a silicon compound at the substrate surface. Silicon tetrachloride ($SiCl_4$) and silane (SiH_4) are most commonly used, although there is some interest in the intermediate compounds, trichlorosilane ($SiHCl_3$) and dichlorosilane (SiH_2Cl_2), as well. The overall reaction for the silicon tetrachloride process is

$$SiCl_4 + 2H_2 \overset{800°C}{\rightleftharpoons} Si + 4HCl$$

It is important to note that this is a reversible reaction, whose end product is either the growth of a silicon layer or the etching of the substrate. The exact nature of this reaction is a function of the mole fraction of $SiCl_4$ in hydrogen.

The silicon tetrachloride reaction is a heterogeneous one, and it occurs on the surface of the silicon, not in the gas phase. Premature nucleation of silicon atoms in the gas phase is thus avoided, resulting in single crystalline growth on the silicon substrate.

One serious problem with the $SiCl_4$ process, autodoping or etchback, comes about because of the reversibility of the reaction. Some silicon and dopant atoms are removed from the layer during growth as a result of the etching reaction. Their rate of removal is roughly proportional to their concentration in the solid phase. The resulting silicon and dopant in the gas phase mix with the incoming gas ambient and modify its impurity concentration, resulting in the deposition of silicon and dopant in a ratio that is proportional to the new concentrations in the gas phase. This redistribution process causes the doping concentration of the epitaxial layer to vary during its growth until an equilibrium situation is reached for layers of sufficient thickness. In addition, it is extremely difficult to grow layers with resistivity in excess of 10 ohm-cm (n-type) by this process.

The overall reaction for the silane process is

$$2SiH_4 \overset{1000°C}{\rightarrow} 2Si + 2H_2$$

Unlike the situation with $SiCl_4$, this is not a reversible reaction. As a result, autodoping does not occur, and resistivities of 50–100 ohm-cm can be achieved by this process. Furthermore, the complete absence of hydrogen chloride gas as an end product makes elaborate venting unnecessary. An additional advantage is that the process can be performed at relatively low temperatures, thus avoiding many of the out-diffusion problems that arise with the $SiCl_4$ process. Its chief disadvantage is the extreme difficulty of

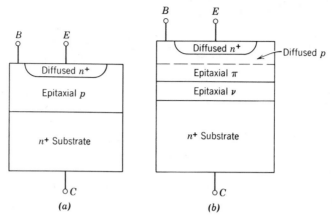

Fig. 6.13   Structures with epitaxially grown junctions.

preventing the gas phase nucleation of silicon particles when this process is used. Thus it usually results in polycrystalline growth, unless considerable attention is paid to the gas flow conditions in the reactor.

Epitaxially grown junctions find their chief application in the fabrication of power transistors. Here the process is often used to form a p-type layer on an n^+-substrate, as the base layer of a single diffused n^+-p-n^+ epitaxial transistor in Fig. 6.13a. A second example is the formation of an ν-type layer, followed by a π-type layer, for a double epitaxial transistor [17], as in Fig. 6.13b (see Section 4.1).

6.3   THE CONTROL OF DEEP LEVELS

The presence of deep levels in silicon has two important consequences. First, they affect the minority carrier lifetime, hence degrade the forward conduction characteristics of thyristors and p-i-n diodes. As a result, special attention must be taken to minimize their presence in such devices. By the same token, their deliberate introduction produces devices with improved recovery characteristics. Thus the optimum lifetime in a thyristor is often a compromise based on the desired operational characteristics, indicating that lifetime control must be relatively precise for these devices. An additional complication results because control must be accomplished without altering the active carrier concentration. This is extremely difficult to do in regions where the initial background concentration is low. In contrast, the situation with low voltage digital circuits is relatively straightforward. Here the usual requirement is to reduce the lifetime as much as possible, without changing the resistivity of the collector material. Since this region is heavily doped to start with (typically $3 \times 10^{16}/cm^3$ in digital microcircuits), lifetime control is a comparatively simple matter.

Almost every chemical impurity outside of columns 3 and 5 of the periodic table exhibits one or more deep impurity levels in silicon [18]. Many of these are found to be present as contaminants in as-grown silicon (Table 6.1). Yet others are introduced into silicon as trace impurities in the chemical solutions and transport gases. Some impurities are present in large quantities in the air, in glassware, in diffusion tube liners, and in the quartz diffusion tubes themselves. The amount of these impurities that is incorporated into the silicon depends on the processing time and temperature involved and on the diffusion-induced stress. Thus contamination problems are most severe with deep-diffused, high voltage structures.

Almost all deep impurities diffuse primarily by an interstitial mechanism and move rapidly through the silicon lattice at processing temperatures. Their diffusion constant is typically 5–6 orders of magnitude higher than that of substitutional impurities such as boron or phosphorus. During processing, they tend to condense around dislocations so as to form metallic precipitates. These precipitates can cause distortion of the potential lines if they are located in the depletion layer of a *p-n* junction, leading to localized regions of high electric field through which excess leakage current flows. Thus a second consequence of deep impurities is that they give rise to "soft" spots and "soft" breakdown characteristics with excess reverse currents at voltages below their avalanche voltage. The elimination or reduction of these precipitates leads to improvement in the sharpness of the reverse characteristic, as well as to improvement in the minority carrier lifetime of the material.

The impurities of special concern in silicon processing are copper, iron, and gold. All diffuse rapidly through the silicon lattice, and have reasonably large solid solubilities and capture cross sections. In contrast, other impurities tend to form inactive silicides and to have extremely low values of solid solubility. Figure 6.6 gives the solid solubilities of these impurities in silicon.

Oxygen is a special case of a deep impurity in silicon. It has a solid solubility of 10^{18} atoms/cm^3 and is often present at this concentration level in CZ material. The oxygen level of FZ material, which is about 1–2 decades lower, is still quite significant.

Active oxygen exhibits deep, donor-like behavior in silicon. With high temperature processing, however, it almost entirely converts to inactive silicon dioxide, which tends to nucleate heterogeneously at sites of vacancy clusters [3]. Typically it precipitates into clumps, 1 μm in diameter, which can act as soft spots and give rise to soft junction characteristics.

Deep levels can also be present as a result of defects in the crystal lattice that either are present in the as-grown material or are created by the subsequent processing. Thus vacancies result in dangling (or unfilled) bonds and are acceptor-like in behavior. They exhibit a deep impurity level at about 0.2 eV from the valence band edge; a second deep acceptor level has also been observed in some situations. Interstitial defects have valence electrons that are not involved in covalent binding with other lattice atoms and exhibit donor-like behavior, with an energy level at approximately 0.2 eV from the conduction band edge. Dislocations tend to behave in much the same way as vacancies, but on a more gross scale. Their observed behavior is also acceptorlike and may be explained by the presence of dangling bonds at the edge of the half-plane comprising the dislocation.

The primary role of defects such as dislocations is in providing sites for the coalescence of deep impurities to form precipitates. Thus they lead to the formation of junctions with soft reverse breakdown characteristics.

6.3.1 Lifetime Improvement Techniques

The most straightforward approach is to use clean processing, to minimize deterioration of the original lifetime of the as-grown material, which can be as high as 1000 μsec. Sealed tube diffusions, for example, are almost exclusively employed for long, high temperature processes such as the formation of deep p-diffusions for thyristors. In addition, the use of symmetrical diffusions on both sides of the slice minimizes the stress created in the vicinity of these junctions (see Section 6.2.1) and inhibits the formation of metallic precipitates. Deep diffusions of this type invariably result in junctions with long lifetimes and hard reverse characteristics.

An interesting observation made in connection with sealed tube diffusions is that subsequent processing in an oxygen atmosphere, at temperatures as low as 700°C, softens the reverse characteristic! On the other hand, the "hard" characteristic can be reestablished if the device is reheated in an evacuated chamber. This can be explained by noting that sealed tube diffusions are conducted in an oxygen-free atmosphere. Silicon dioxide clusters are highly mobile at diffusion temperatures, and they migrate to the surface of the silicon because of the extremely low partial pressure of oxygen in the sealed tube. Thus they move away from the space charge region where they can give rise to soft junction characteristics. Additionally, it is felt that the migration of oxygen to the surface provides the driving force for the movement of fast diffusing impurities, away from the junction region [19].

The deliberate manipulation of clusters and deep impurities *away* from the space charge regions thus presents a powerful alternate technique for their reduction, by a process known as *gettering*. A number of gettering procedures currently in use are discussed in the following sections.

6.3.1.1 Halogenic Dopant Sources

The problems of contamination are particularly severe in open tube diffusion systems, where extreme care must be taken to avoid contamination with impurities other than those intentionally used as dopants. Some of these impurities are already present in the starting silicon; yet others are present in the carrier gases or dopant source, and some exist in the atmosphere and enter the diffusion tube from its open end. Typically, all exhibit deep levels.

The use of halogenic dopant compounds such as $POCl_3$, PBr_3, and BBr_3 is found to improve this situation. In effect, the halogen, liberated during diffusion, reacts with any impurities in the gas stream, as well as with impurities within the silicon that reach the exposed surface during their rapid movement at diffusion temperatures. This reaction converts them to their more volatile halides, which leave the system by incorporation into the gas stream. Thus the use of these dopant sources getters the silicon during this process. Halogenic dopant sources must be used with care, however, since the liberation of excess bromine or chlorine can result in local dissolution of the silicon itself. This phenomenon, known as *halogen pitting*, can destroy the flatness of the silicon surface and cause difficulty in subsequent processing.

Halogens are also used to provide a gettering action during the thermal oxidation of silicon. This is done by premixing the oxygen with anhydrous hydrogen chloride gas. Recently trichloroethylene has been used as the halogen source in this process.

6.3.1.2 Metallic Gettering

To achieve metallic gettering, one evaporates a metal such as nickel or zinc on the silicon slice, which is then heated in an inert atmosphere (typically 1 hr at 1100°C). Impurities such as gold tend to move preferentially into this layer because of its higher solubility, resulting in an improvement in the lifetime. Metallic getters have not been found to be very effective; in addition, they often cause gross damage to the silicon surface.

6.3.1.3 Mechanical Damage

Mechanical damage of the silicon surface, by sandblasting or abrasion, results in the formation of an almost infinite supply of vacancies. During high temperature processing, these vacancies act as sinks for silicon dioxide clusters and fast-moving deep impurities that diffuse throughout the slice.

Both copper and iron diffuse rapidly, by an interstitial mechanism, but ultimately take up substitutional sites by combining with vacancies. Moreover, since substitutional copper and iron have a lower solid solubility than the interstitial, these materials freeze out upon cooling and tend to condense around dislocations. Once there, however, they may either remain in the slice or leave by out-diffusion. The vapor pressure of copper is 15–20 times higher than that of iron or gold. Consequently nearly all this impurity leaves by out-diffusion. On the other hand, there is little out-diffusion with iron; typically, about 33% segregates in this layer of dislocations, the rest being retained in the bulk.

Gold moves primarily by an interstitial mechanism and also takes up substitutional sites in the lattice by combining with vacancies. Unlike copper and iron, the solid solubility of substitutional gold is much higher than that of the interstitial species; thus it remains in solution upon cooling. Furthermore, since gold does not form compounds with silicon, it does not segregate by this process.

Controlled amounts of damage can also be introduced into a silicon slice by the use of a deposited layer of silicon nitride [20]. Here the high interfacial stress created by the deposition process results in formation of a dislocation network on that face of the silicon toward which it is desired that deep impurities migrate. Ion implantation is also being explored as an alternate technique [21] for producing the desired surface damage.

6.3.1.4 Glassy Layers

Both borosilicate (BSG) and phosphosilicate (PSG) glasses, in contact with the silicon at elevated temperature, are extremely effective in improving the properties of the bulk silicon. A theory [22] that has been proposed to explain the effectiveness of this gettering process is that heat treatment of the sample results in creation of a diffusion-induced stress well ahead of the diffusing front. Typically, a 1 hr heat treatment at 1200°C results in a 5–10 μm region of stress. This is outside the space charge region of deep-diffused, high voltage junctions, and it can serve as a sink for impurities.

Out-diffusion plays an important role with copper [23], and about two-thirds of this material leaves the silicon slice, while the rest piles up in the damaged region *under* the glass. Very little out-diffusion occurs with elemental iron. On the other hand, it appears that compounds are formed between the iron and the glass, with about one-third lost by incorporation into the glass.

Out-diffusion is almost negligible with gold, and all this material remains within the silicon slice. Typically, 20% is located in the glass after heat treatment, with the rest divided between the bulk of the silicon and a 0–10 μm layer under the glass. About 60% of the gold is collected *under the glass*, and the remaining 20% is in the bulk silicon when PSG is used. With BSG, however, as much as 40% remains in the bulk regions. The superiority of PSG as a getter for gold is possibly attributable to the enhanced solid solubility of substitutional gold in silicon in the presence of phosphorus, due to ion pairing [24]. On the other hand, BSG results in boron diffusion into silicon, which getters gold by the formation of compounds that precipitate at moderate to heavy contamination levels. However it is less effective for dilute gold concentrations.

Glasses used for gettering are usually formed by the oxidation of the common halogenic dopant sources. Thus BBr_3 is used for the formation of BSG, whereas $POCl_3$ and PBr_3 have been used for PSG-gettering. The peak concentration for gold is usually found to be on the surface for slices that are PBr_3-gettered, but somewhat deeper when $POCl_3$ is used. This is probably because the dislocation network configurations obtained from these two glasses are comparably different [25].

6.3.2 Lifetime Reduction Techniques

The reduction of lifetime in a controllable manner is particularly important in fast thyristors, since the recovery time is directly related to the minority carrier lifetime in the base regions. Chemical defects (dopants) as well as crystal damage can be used to this end. In both cases, however, the reduction of lifetime brings with it all the harmful side effects attributed to the presence of deep levels. Thus they are fast diffusers and cannot be introduced selectively into the device. They go everywhere, with higher localized concentration in regions of high background doping. In addition, they tend to segregate around dislocations and result in junctions with soft reverse characteristics. Often these precipitates act as initiation sites for second breakdown phenomena.

6.3.2.1 Doping

The most common technique for the reduction of lifetime consists of the deliberate introduction of controlled amounts of impurities, which create deep levels in the silicon lattice. The two impurities under active study for this application are gold and platinum. Figure 6.14 shows the energy levels [26, 27] associated with these impurities. Their electronic properties are listed in Table 6.3 [28–30]. A comparison of their properties follows.

Table 6.3   Capture Rate Constants for Gold and Platinum in Silicon

	Capture Rate Constants (cm^3/sec)	
	Gold [28]	Platinum [30]
Electron capture at E_1^+	6.3×10^{-8}	2.2×10^{-8} (Ref. 29)
Hole capture at E_1^+	2.4×10^{-8}	1.2×10^{-9}
Electron capture at E_1^-	1.65×10^{-9}	2.4×10^{-9}
Hole capture at E_1^-	1.15×10^{-7}	1.5×10^{-7} (Ref. 29)
Electron capture at E_2^-	—	3.2×10^{-7}
Hole capture at E_2^-	—	2.7×10^{-5}

1. Gold forms no compounds in silicon. Thus it is readily introduced into silicon by using an evaporated layer of the element as a source. Platinum, on the other hand, forms numerous silicides, and its actual incorporation into the silicon from the element is highly variable. Recently it has been found that when extremely small amounts of platinum are used as a source, the results are highly reproducible. One technique is to make a slurry of a platinum compound such as the oxide in a carrier liquid (e.g., polyvinyl alcohol) and to paint or spin this on the silicon surface.
2. Both gold and platinum diffuse rapidly into silicon. In either case, about 90% of these impurities occupies electronically active substitutional sites. The rest of the gold is inactive and is assumed to be in

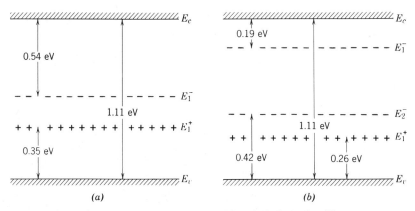

Fig. 6.14   The energy levels of gold and platinum in silicon.

interstitial sites. In contrast, the remaining platinum is electronically active and is considered to give rise to the acceptor level at 0.42 eV above the valence band (see Fig. 6.14). The capture cross section of this level is extremely large and accounts for most of the lifetime reduction properties of this dopant [30].

3. The impurity levels of platinum are highly asymmetric, whereas one gold level (the acceptor) is almost in the center of the energy gap. This gives rise to different lifetime characteristics [31] for the two materials. In particular, the space charge generation lifetime in platinum at room temperature (see Section 1.1) is many hundred times larger than its low level lifetime. Even at 125°C, the space charge generation lifetime in platinum is 20 times larger. As a result, platinum doping can reduce the low level lifetime in the neutral regions of a diode without a comparable increase in its leakage current. In contrast, a direct consequence of the symmetrical location of the gold acceptor level is that leakage current is inversely related to the lifetime in its neutral regions. The position of the energy levels also affects the relationship between high and low level lifetimes.

Gold is an extremely rapid diffuser in silicon, with a diffusion coefficient about 5 orders of magnitude larger than that of boron or phosphorus. Usually the last impurity to be diffused into the device, it diffuses into the silicon lattice by two mechanisms: a fast interstitial mechanism with $D_i \cong 10^{-5}$ cm^2/sec at 1050°C, and a slow substitutional process with $D_s \cong 10^{-10}$ cm^2/sec at 1050°C. In addition, the concentrations of gold in interstitial and substitutional sites (defined as c_i and c_s, respectively) are quite different and are a function of diffusion temperature (e.g., c_i is about $0.1c_s$ at 1050°C).

On introduction into the silicon lattice, interstitially located gold atoms can become substitutional by dropping into vacant lattice sites when they come within their capture range. This upsets the equilibrium concentration of these vacancies, and more are thermally generated. In silicon having a high defect concentration, this readily occurs in the vicinity of a dislocation and results in the dislocation's climb.* As a consequence, the concentration of vacant lattice sites tends to preserve its equilibrium value.

In silicon having a low defect concentration, however, these vacant lattice sites must be provided from the surface of the wafer. Since this is a much slower process, their number may become less than the equilibrium value; the diffusion rate now depends on a dissociative mechanism

---

*The presence of oxygen in the silicon tends to pin the dislocations, hence is an additional complicating factor in the diffusion of gold.

whereby gold in lattice sites moves into interstitial sites, leaving behind vacancies. As a result of these processes, the diffusion of gold in silicon is a strong function of the defect concentration for that particular crystal, in addition to the temperature.

Gold diffusion is usually carried out from a 100–500 Å thick, vacuum-evaporated layer on the silicon. Diffusion proceeds from a liquid gold-silicon alloy, as indicated by the phase diagram of Fig. 6.10, and results in silicon damage to a depth of many microns. Diffusion is normally carried out in the 800–1050°C temperature range. The concentration is controlled by means of this temperature, and the time is chosen to be more than sufficient to cause the gold to diffuse through the entire device. The precise diffusion temperature is selected on a cut-and-try basis and is altered to adjust the gold concentration in the silicon; however the actual concentration is also a function of such variables as the manner in which the wafer is cooled to room temperature. The repeatability of this quenching cycle is important in ensuring consistent results because out-diffusion effects can be significant. The choice of the precise diffusion time and temperature is thus determined by trial and error, with a desired end effect in mind.

Platinum diffusions are usually conducted from platinum compounds that are incorporated into binders and used as spin-on dopants. A colloidal suspension of a platinum compound in silicon dioxide is also available commercially for this purpose.* The technology for this dopant is quite recent, and we do not know how it diffuses into the silicon lattice.

6.3.2.2 Electron Irradiation

The introduction of deep levels by electron bombardment represents a new and highly promising approach to the control of lifetime in power semiconductors, especially thyristors, where junction temperatures ordinarily do not exceed 150°C. Here uncapped but otherwise fully packaged devices are subjected to electron bombardment at room temperature, followed by a low temperature anneal, until the desired operating characteristics are obtained.

During electron bombardment [32] much of the energy of the incident particles goes into transient electronic processes such as ionization and excitation. The remainder, however, goes into atomic process such as the dislodging of atoms from their sites to form defects. Many defects formed in this way are unstable and anneal out at room temperatures. However defects such as divacancies anneal out at 225–290°C, thus are stable during

---

*Platinumsilicafilm, manufactured by the Emulsitone Co., Milburn, N.J.

the operation of thyristors. A single 2 MeV electron/cm^2 typically produces about 0.26 stable defects per cubic centimeter.

The role of defects is to introduce deep levels, thus to reduce the minority carrier lifetime. Thus,

$$\frac{1}{\tau} = \frac{1}{\tau_0} + K'\phi_e \qquad (6.3a)$$

where τ and τ_0 are minority carrier lifetimes after and before radiation, ϕ_e is the radiation fluence per square centimeter, and K' is a damage factor. For high resistivity n-type silicon and 2 MeV electrons, this equation has been empirically fitted [33] to

$$\frac{1}{\tau} = 0.029 + 8.2 \times 10^{-15}\phi_e \qquad (6.3b)$$

for $\phi_e \leqslant 7 \times 10^{13}/cm^2$ and τ in microseconds, and to

$$\frac{1}{\tau} = 0.452 + 2.19 \times 10^{-15}\phi_e \qquad (6.3c)$$

for electron fluences between $7 \times 10^{13}/cm^2$ and $2 \times 10^{14}/cm^2$. For example, a dose of 7×10^{13} electrons/cm^2 in the 2 MeV range typically reduces the minority carrier lifetime in the n-base of a thyristor to 1.6 μsec (i.e., a reduction by a factor of 21 for a thyristor with an unradiated lifetime of 35 μsec). This can result in significant improvement of the recovery characteristics of a thyristor designed for fast switching applications.

Electron irradiation [34] gives rise to four energy levels, at $E_v + 0.27$ eV, $E_c - 0.17$ eV, $E_c - 0.23$ eV, and $E_c - 0.41$ eV. Annealing for 36 hr at 300°C alters the defect structure, with the last two levels converting to one level at $E_c - 0.36$ eV. Lifetime reduction is not accompanied by a significant increase in the leakage current because of the asymmetrical nature of this energy level, which dominates the generation and recombination process. The space charge generation lifetime is typically about 20 times that in the neutral region at 125°C, and many hundred times larger at room temperature.

Electron irradiation has a number of advantages over impurity doping for the control of lifetime. First, excellent process reproducibility is obtained, since dosage can be accurately monitored. Next, irradiation is performed at room temperature on pretested but uncapped devices, not at high temperatures. Finally, electron irradiation effects are relatively uniform throughout the device. In contrast, impurity doping is a high temperature process and always results in some clustering around dislocations, with an attendant increase in soft spot formation. Given these advantages, it is highly probable that electron irradiation will be increasingly adopted for the manufacture of fast recovery thyristors.

6.4 OHMIC CONTACTS

In addition to excellent mechanical properties, an ohmic contact must have low electrical resistance and an essentially linear voltage-current characteristic. Furthermore, it must serve purely as a means for getting current into and out of the semiconductor while playing no part in the active processes occurring within the device.

An ohmic contact can readily be achieved to a heavily doped, degenerate semiconductor, since this material has an extremely short minority carrier lifetime. Thus a good, strong mechanical connection is the principal consideration when making such a contact. Regions of this type are often directly attached to the leads, or to the case to which the circuit connection is to be made.

An ohmic contact to a lightly doped semiconductor is accomplished by forming an n^+-contact to an n-type region or a p^+-contact to a p-type region. These high-low junctions [35], of the n^+-n or p^+-p type, have extremely high leakage currents that completely mask the usual diode-like behavior. Here their voltage-current characteristics approximate a straight line going through the origin and are essentially "ohmic."

Techniques for junction formation described in Section 6.2 can also be used for the fabrication of ohmic contacts. A few examples follow.

1. The diffusion of an n^+-region into the n-collector of a microcircuit transistor (Fig. 6.15a), followed by a p^+-alloyed interconnection layer of aluminum. This combination results in an n-n^+ junction followed by an n^+-p^+ junction in series with it.
2. The alloying of an aluminum p^+-region to the p-gate region of a thyristor (Fig. 6.15b).
3. The alloying of an aluminum p^+-region to the p-anode region of a thyristor, to form a p^+-p contact (Fig. 6.15b).
4. The epitaxial growth of an n-collector region to an n^+-substrate, as shown in Fig. 6.15c for a transistor. Here the n^+-region provides for ohmic contact to the n-collector, even though the latter is epitaxially grown on the former.

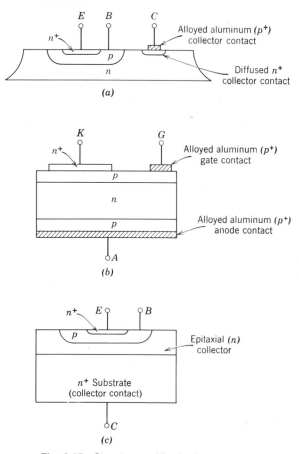

Fig. 6.15 Structures with ohmic contacts.

6.5 LEAD AND DIE ATTACHMENT

Semiconductor power devices have ratings based primarily on thermal considerations. As a consequence, their thermal design determines their power dissipation capability, as well as their ability to control power in different circuit situations. The latter problem has been extensively treated in the applications literature [36–38]. Our emphasis is on the basic aspects of thermal design and of attachment systems that can operate under fluctuating load conditions.

6.5.1 Thermal Considerations

Power devices may fail catastrophically if the junction temperature becomes high enough to cause thermal runaway and melting. A much lower functional limit is set by temperature increases that result in changes in device characteristics (e.g., the forward breakover voltage or the recovery time) and failure to meet device specifications.

Heat generation occurs primarily within the volume of the semiconductor pellet. This heat must be removed as efficiently as possible by some form of thermal exchange with the ambient, by the processes of conduction, convection, or radiation. Heat loss to the case and heat sink is primarily by conduction. Heat loss by radiation accounts for only 1–2% of the total and can be ignored in most situations. Finally, loss from the heat sink to the air is primarily by convection. When liquid cooling is used, the heat loss is by conduction to the liquid medium through the walls of the heat exchanger.

Heat transfer by conduction is conveniently described by means of an electrical analog. Thus if any thermal element has a temperature differential of ΔT across its ends, then

$$\Delta T = PR_\theta \qquad (6.4)$$

where P is the power dissipated in the element (W), and R_θ is its thermal resistance (°C/W). Thermal resistance is related to thermal resistivity in the same manner as its electrical counterpart. If L is the length of the heat flow path (cm), A is its cross-sectional area (cm²), and K is its thermal conductivity (W/cm-°C), then

$$R_\theta = \frac{1}{K} \frac{L}{A} \qquad (6.5)$$

Values of the thermal resistivity $1/K$ are given in Table 6.4 for a number of materials of interest to designers of semiconductor power devices.

A complete cooling path, from source of heat to the ambient, can be considered to be the series connection of separate elements that make up this path. For example, the temperature rise of a thermal system consisting of junction, die bond, case, sink, and ambient is given by

$$\Delta T_{JA} = PR_{\theta JA} \qquad (6.6)$$

where

$$R_{\theta JA} = R_{\theta JD} + R_{\theta DC} + R_{\theta CS} + R_{\theta SA} \qquad (6.7)$$

and the subscripts J, D, C, S, and A refer to junction, die bond, case, sink, and ambient, respectively. In some thermal systems, more than one path is

Table 6.4   Thermal and Electrical Properties of Selected Materials

Material	Thermal Expansion Coefficient $(°C)^{-1} \times 10^{-6}$	Thermal Resistivity (°C-cm/W)	Volume Heat Capacity (W-sec)/°C-cm³)
Si	2.6	0.69	1.75
Mo	5.4	0.66	2.75
W	4.5	0.6	2.75
Al	23.1	0.43	2.6
Au	14.2	0.34	2.5
Ag	19.6	0.24	1.81
Cu	17.3	0.26	3.37
Kovar	4.6–5.2	5–6.1	3.7–5.4
Mild steel	15.1	1.65	3.3–3.6
Thermal grease	—	150	—
Mica	9–13	200–300	0.54–0.66
Al_2O_3	6.6	4.0	3.31
BeO	6.5	0.6	3.68

available for heat removal to the ambient. Here, thermal analysis can be carried out by considering parallel combinations of thermal resistance. This occurs in high power thyristors where both sides of the device are used for heat removal—an important situation.

It is often necessary to electrically insulate a power device from its heat sink. This is usually done by means of a thin mica spacer that is lubricated with a thermally conductive grease. Sometimes an aluminum heat sink is used with an insulating layer of anodized aluminum oxide. In either case, the thermal resistance associated with these separate components must also be included in the analysis, and it is a sensitive function of surface preparation and the torque with which the device is bolted to the heat sink.

6.5.1.1   Transient Thermal Impedance

The concept of thermal resistance can be extended to thermal impedance for time-varying situations. Following the electrical analog, the thermal time constant of an element is $R_\theta C_\theta$, where C_θ is given by

$$C_\theta = C\rho V \qquad (6.8)$$

Here C_θ is the thermal capacitance (W-sec/°C), C is the specific heat (W-sec/g-°C), ρ is the density (g/cm³), and V is the volume (cm³). The product $C\rho$ is referred to as the volume heat capacity, and values are given in Table 6.4 for materials of interest.

Consider what happens when a step function of power is applied to such an element, resulting in a steady state temperature rise of ΔT_F. The temperature rise at any given time can be written as

$$\Delta T(t) = \Delta T_F \left(1 - e^{-t/R_\theta C_\theta}\right) \qquad (6.9)$$

The response of a single element can be extended to a complex system such as a power device, whose thermal equivalent circuit comprises a ladder network of the separate resistance and capacitance terms shown in Fig. 6.16a. The transient response of such a network to a step of input power takes the form of a series of exponential terms; unfortunately, the time constant of each term is a complex function of all the individual time constants and cannot be easily interpreted.

Figure 6.16b gives the temperature response to an input step of power. Here at any time t_0, the temperature rise is given by ΔT_0. A transient thermal impedance $R_{\theta t}(t_0)$ can now be defined for this time, such that

$$\Delta T_0 = PR_{\theta t}(t_0) \qquad (6.10)$$

For large values of time, $R_{\theta t}$ approaches the steady state thermal resistance. For short times, however, heat flow is primarily by diffusion within

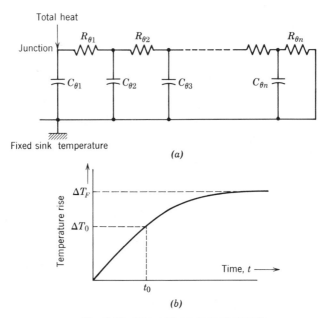

Fig. 6.16 Thermal equivalent circuit.

the silicon pellet, and $R_{\theta t}$ is approximately proportional* to $(t)^{1/2}$. It should be emphasized, however, that this relationship cannot be used with thyristors for time intervals below the plasma propagation time, since this value is based on the assumption that the entire device is in conduction.

Extremely short overloads, of the type that occur under surge or fault conditions, are limited to a few cycles in duration. Here the junction temperature exceeds its maximum rating and all operational parameters are severely affected. However the low transient thermal impedance offered by the device in this region of operation is often sufficient to handle the power that is dissipated.

The transient thermal impedance of a typical thyristor device mounted on an infinite heat sink (case temperature fixed), is shown in Fig. 6.17. Also shown is the device mounted on a finite heat sink, such as a copper fin. This results in an additional $R_\theta C_\theta$ term to be included in the thermal analog. However the time constant of such a sink is considerably larger than the other time constants, dominating the thermal impedance characteristic for longer time intervals, as Fig. 6.17 reveals.

Transient thermal impedance data, derived on the basis of a step input of power, can be used to calculate the thermal response of devices for a

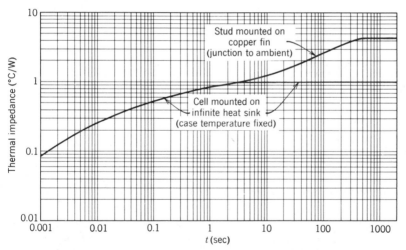

Fig. 6.17 Thermal impedance of a typical thyristor. From Grafham and Hey, [37]. Reprinted by permission of the General Electric Company, Semiconductor Products Department, Auburn, N. Y.

*A similar relationship was obtained in a study of the initiation time for second breakdown (see Section 1.5.2.2), for pulse durations where diffusional heat flow processes are significant.

variety of one-shot and repetitive pulse inputs. Computations for commonly encountered situations have been tabulated [36–38] and are of great value to the circuit designer who must specify a device and its derating characteristics. However the steady state thermal resistance and capacitance terms are of primary interest in understanding device operation and in device design.

6.5.2 Solder Systems

In many situations, semiconductor power devices are subjected to thermal cycling during their operation; often they require to survive as many as 500 thermal cycles at a case temperature of 65°C, during the period of their useful life [39]. Both lead and die attachment systems must be designed to maintain their integrity during this type of service. *Hard* and *soft* soldering are commonly used techniques for these systems. Soft solder is employed for die attachment in small devices, in devices for consumer markets, and in power transistors. The most commonly used soft solders consist of the lead-tin alloys, a typical composition being 95% lead and 5% tin by weight, and having a melting point of 310–314°C (Fig. 6.18). Alloys of lead-indium-silver are also used in some situations. One example is an alloy of 92.5% lead, 5% indium, and 2.5% silver, which is sometimes used because of its better thermal cycling characteristics.

The stress created during thermal cycling, by the differential thermal expansion of the silicon pellet and the case, is taken up by the formation of dislocations. These tend to pile up at preferred sites such as voids, ultimately leading to crack formation and failure by *thermal fatigue*. With soft solders, the cyclic stress is taken up by plastic deformation of the interface layer. This material tends to anneal partially at some critical temperature,* permitting a return to its normal unstressed condition. Thus soft solders can perform satisfactorily under cyclic conditions, if the solder system is exposed to temperatures in excess of some critical value during device operation.

Considerable care must be taken to prevent void formation in the soft solder attachment of devices. These voids lead to nonuniformity in heat sinking [40], which affects the current distribution in the device. With transistors, this causes current crowding; second breakdown has often been found to occur in the vicinity of such voids.

Soft solders cannot be used in large area devices, or in devices that are

---

*This critical temperature, in degrees Kelvin, is approximately equal to two-thirds the melting point in degrees Kelvin.

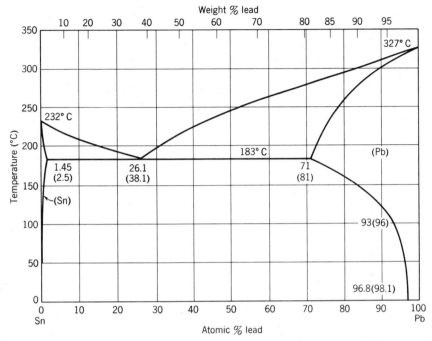

Fig. 6.18 The lead-tin system. From *Constitution of Binary Alloys*, by M. Hansen and A. Anderko [15]. Copyright 1958. Used with permission of the McGraw-Hill Book Company.

subjected to excessive thermal cycling. Here, hard solders such as gold-silicon and aluminum-silicon are used throughout, and the differential thermal expansion stress is taken up by the solder-interface layer, or by means of a disc of intermediate material placed between the pellet and the case. Ideally, this intermediate disc should have a close match to silicon in its thermal expansion characteristics, to relieve the shear stress at the (somewhat fragile) silicon face and transfer it to the more sturdy disc-case combination. Both tungsten and molybdenum are suitable choices (see Table 6.4) for this application, and are close to silicon in thermal expansion characteristics. Furthermore, both are relatively immune to fatigue under cyclic stress conditions. Tungsten has a slightly lower thermal resistivity than molybdenum and is a better thermal match to silicon. Consequently it is the preferred metal for this intermediate disc in high power applications. A schematization of such a device (Fig. 6.19) illustrates the different components of the lead and die attachment.

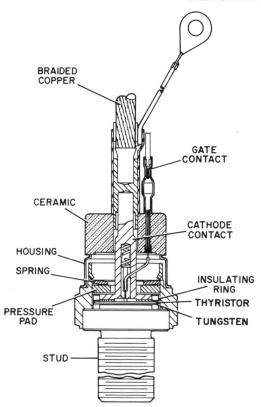

Fig. 6.19 Schematic of a stud-mounted thyristor. Courtesy of the General Electric Company, Semiconductor Products Department, Auburn, New York.

6.5.3   Compression Systems

Compression attachment is an alternate technique used for high power device structures. Here, both thermal and electrical connections are made by the application of a compressive force to copper terminals on either side of the device. Spring washers, often employed for this purpose, hold the silicon pellet within its package.

For large devices, 33 mm or more in diameter, it is customary to directly clamp the silicon slice between copper pole pieces, which serve as electrical terminals. Thin ($\cong 125$ μm) shims of copper or silver foil are usually interposed between the silicon slice and the pole pieces to aid in making intimate electrical and thermal contact over these large area surfaces, and to accommodate for any surface roughness that may exist. A clamping pressure of 100–150 kg/cm² is commonly used. This has little effect on the silicon, which can stand compressive stresses in excess of 2800 kg/cm² without physical damage. Pole pieces may be cooled by forced air or water as the need arises, with cooling being provided on both sides. Figure 6.20 is a schematic diagram for a compression system of this type.

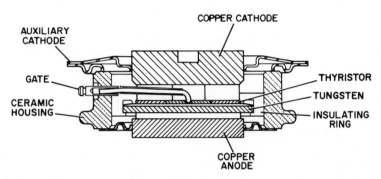

Fig. 6.20 Schematic of a compression-mounted thyristor. Courtesy of the General Electric Company, Semiconductor Products Department, Auburn, New York.

Table 6.5 gives values of thermal resistance [41] existing between the junction and the heat sink ($R_{\theta JS}$), as well as between the heat sink and the cooling medium ($R_{\theta S\text{-air}}$ or $R_{\theta S\text{-water}}$). As expected, the use of water as a cooling medium results in an improvement over forced air. It is interesting to note from this table, however, that the thermal resistance of the junction to the heat sink is the limiting factor in devices less than 50 mm in diameter. With larger devices, the ability to remove heat by way of the cooling medium sets the ultimate limit to the power they can dissipate. Thus still larger devices will require more efficient means for heat removal, such as heat pipes.

Table 6.5 Thermal Resistances for Large Area Power Devices [41]

Thermal Resistance	Nominal Diameter (mm)			
(°C/Watt)	33	50	75	100
$R_{\theta JS}$	0.08	0.04	0.02	0.015
$R_{\theta S\text{-air}}$	0.05	0.04	0.035	0.03
$R_{\theta S\text{-water}}$	0.02	0.02	0.015	0.015

6.6 SURFACE PASSIVATION

The junctions of power devices are generally coated with a protective layer before encapsulation. The type of coating or surface passivation employed depends on the intended end use for the devices. Small, low voltage devices, especially those designed for the low cost, high volume market, are normally made by planar processing, which result in the formation of a thin, thermally grown oxide ($\cong 0.5$ μm thick) on the silicon surface. Although this provides some degree of protection, an additional coat of a phosphosilicate glass is often used. This glass is formed by the simultaneous decomposition of silane and phosphine in oxygen [42], at temperatures on the order of 300–400°C. Its thermal expansion characteristics can be adjusted by varying the P_2O_5 content, and this permits relatively thick layers (1–10 μm) to be grown without cracking.

6.6.1 Organic Coatings and Varnishes

It has been shown that extremely high voltage devices can be fabricated by surface contouring techniques (see Section 2.6). Even though the peak electric field is located internally, surface leakage current is significantly reduced by the application of coatings to the bevel region. These coatings typically consist [43] of methyl phenyl compounds in toluene or xylene, or of silicone rubbers. The coatings are applied to the bevel region, often with catalyzers, and cured at an elevated temperature.

Surface coatings of this type have many disadvantages. For example, their electronic properties are not well understood, and chemically identical materials give very different results from batch to batch. In addition, the acceptance (or rejection) of a device cannot be established until after the coating has been applied and cured, and the device encapsulated. Thus devices are often rejected after a number of expensive processing steps have been performed on them.

6.6.2 Glasses

Recently a variety of glasses have been developed in an effort to avoid the problems just outlined. For example, Innotech Type 760* is a lead-alumina-borosilicate glass having a thermal expansion coefficient of 4.8×10^{-6}/°C. Another glass that is suitable for device encapsulation is GE Type 351,† which has a thermal expansion coefficient of 4.45×10^{-6}/°C. This glass is primarily of a zinc-borosilicate composition [44], with small

---

*Innotech Corp., Norwalk, Conn.
†General Electric Co., Schenectady, N.Y.

amounts of Bi_2O_3, CeO_2, Sb_2O_3, and PbO incorporated into it to alter its crystallization characteristics. Both these glasses can be readily deposited in 25 μm thick layers, which are useful in high voltage structures. They have two advantages over organic junction coatings in that they are more predictable in their characteristics and more reproducible from batch to batch.

Devices can be coated with glass in a number of ways. One of the simplest is to apply a slurry of the glass in an organic binder (a butyl carbitol–ethyl cellulose mixture) to the silicon surface, and to scrape it into a uniform layer by means of a flat blade (the so-called doctor-blade technique). This layer is fired at 700°C to form a continuous glass coating, after a preliminary bake-out to drive off the binder.

A second method is to place the silicon slice in a container in which the glass is suspended in a liquid. Spinning of this container at high speed results in the centrifugal sedimentation of the glass on the silicon surface. Again, bake-out and subsequent firing are used to form the glass layer.

A third approach is to coat the silicon surface by electrophoretic methods [45]. The silicon slice is made the cathode of an electrophoretic cell in which a noble metal such as gold is used as the anode. A colloidal suspension of the glass in a medium such as acetone, ethyl acetate, or isopropyl alcohol is used in this cell. A small amount of ammonia or hydrofluoric acid is also added to achieve uniform glass deposition. In operation, the glass particles become charged and migrate to the silicon surface, where they are deposited. The slice is subsequently fired to form the glass coating.

A key advantage of the electrophoretic process is that by suitable masking, glass can be formed on selective regions of a semiconductor device, in addition to following its physical contours. This characteristic has been exploited in the fabrication of large numbers of high voltage transistors on a single silicon slice. For these devices, deep "mesa" etching is used to establish a bevel at the collector-base junctions. The electrophoresis technique allows the formation of a uniformly thick glass layer over these junctions, making it possible to test devices for reverse breakdown at the wafer level before dicing. By this means, chips that might fail the subsequent reverse breakdown test are detected before bonding and packaging.

6.6.3 Thermal Oxides

The most straightforward approach to surface passivation is to thermally grow either doped or undoped silicon dioxide on the surface of the junction. Although universally employed for low voltage structures, this approach has generally been found unsatisfactory for high voltage devices for two reasons. First, the presence of impurities in the oxide softens the otherwise abrupt diode characteristics. Second, this process results in the formation of SiO_2 precipitates or soft spots during the high temperature oxidation step.

Ultra-clean techniques for oxidation at relatively low temperatures have recently been developed for this application [46], and suitable coatings have been obtained for beveled devices with breakdown voltages up to 2.8 kV. These techniques have been applied to the fabrication of high voltage thyristors. After the conventional sealed tube gallium diffusion, the slices are subjected to a borosilicate glass gettering treatment, then beveled to the appropriate angle. Next thermal oxidation is carried out at 800°C in a diffusion tube made of semiconductor grade silicon, rather than of conventional quartz. Oxide layer thickness is typically 600–1000 Å, subsequently capped with a 2000 Å thick deposited layer of silicon nitride.

Success with thermal oxidation is quite recent, and the technique will no doubt be strenuously pursued because of its inherent simplicity. Furthermore, it should ultimately prove superior to other methods, since the oxide is grown directly out of the silicon, not by the application of foreign matter to the surface.

6.6.4 Polycrystalline Silicon Films

The use of thermal oxides is particularly unsuited for high voltage planar junction devices that have high electric fields at the surface because it gives rise to a number of types of instability, including the following:

1. Instabilities due to the movement in the oxide of sodium ions [1], which modulate the conductivity of the underlying material and often lead to surface inversion effects.
2. Trapping of charge in the oxide during avalanche breakdown, resulting in a temporary increase in the breakdown voltage (the so-called walk-out effect [47]).
3. Degradation of the low level current gain in transistors [48] after avalanche breakdown of the emitter-base junction, caused by an increase in the density of fast surface states.

Films of semi-insulating polycrystalline silicon can be used [49] in this application, since they are almost electrically neutral. Furthermore, their electronic properties can be changed by oxygen doping until they are completely converted to SiO_2 (66.7% atomic concentration of oxygen). Experimental data on deep planar junctions reveal that the breakdown voltage falls with increasing oxygen concentration, up to about 8%

(atomic). Concurrently, the resistivity of the films increases as the material shifts toward SiO_2, and the reverse leakage current falls.

The use of a semi-insulating silicon film with transistors results in a degradation of their low level current gain, since it acts as a resistive emitter-base shunt. It has been demonstrated experimentally that this current gain is unaffected for oxygen concentrations above 20%. Thus it is possible to effect a tradeoff on these three device parameters by judicious adjustment of the oxygen content.

Polycrystalline silicon films can also be nitrogen doped until they become silicon nitride, which can be used as a cover layer because of its well-known ability to inhibit the transport of both sodium ions and water to the silicon surface. Here, however, no balance need be struck on the film composition, since best protection is achieved with full nitridation.

The process described here is inherently expensive because of the problems of growing and doping multilayer films on semiconductor surfaces. Nevertheless, stable planar junctions with breakdown voltage in excess of 10 kV have been achieved by this process, and it is under active consideration.

6.7 ENCAPSULATION AND PACKAGING

Hermetic encapsulation of power transistors is usually done by the attachment of the device in a metal case; leads to the device are brought out from the bottom by way of glass-to-metal seals. Power thyristors and rectifiers, which are often subjected to rapid thermal cycling, make extensive use of ceramic packages, and ceramic-to-metal seals, which are more rugged under thermal cycling. Low power, commercial grade devices (transistors, rectifiers, and thyristors) are often directly encapsulated in integrally molded epoxy packages.

A recent innovation is the packaging of a number of medium power semiconductor devices, suitably interconnected, into a hybrid integrated power circuit module. The devices are bonded to an insulating substrate on which a conductive pattern has been previously formed. The formation of a highly conductive metallic pattern on substrate materials such as alumina or beryllia is thus a key step in hybrid integrated circuit fabrication.

Strong bonding between the conductor and substrate is essential. This is usually accomplished by the formation of an intermediate layer, which must have good thermal properties for power applications. In low power, hybrid integrated circuits, conductors of noble metal (e.g., gold and platinum-gold) have been used; however these materials have low electrical conductivity relative to copper and cannot be used at high current levels.

A common approach for power applications is to screen print a mixture of molybdenum and manganese oxides, fire it onto the substrate, and subsequently plate it with copper to a thickness of 25 μm or more to improve its conductivity and current-carrying capacity. This process is relatively complex, and simpler approaches have been sought.

A recent technique that shows [50] considerable promise in this area is the direct bonding of copper to ceramics by a $Cu-Cu_2O$ eutectic bond. Formed at 1065°C, this eutectic consists of 4.68% Cu_2O by weight. Formation of this bond is accomplished by contacting a precut foil of copper to the ceramic substrate and firing at 1070°C in a nitrogen atmosphere with about 0.02–0.08% oxygen. The interfacial energy between the copper and the ceramic drops rapidly in the presence of this oxygen, resulting in a strong, permanent bond with low thermal resistance.

The $Cu-Cu_2O$ bond technique can be used for direct bonding of copper to alumina, beryllia, boron nitride, and zinc oxide substrates. The copper foil can be cut to any thickness, thus it presents minimal electrical resistance in the circuit. Glass-encapsulated devices, both discrete and integrated, can be directly bonded to these conductors by hard or soft solder techniques to form complete subsystems. Finally, since the copper foil is only heated below its melting point (i.e., 1083°C), it remains intact and can be extended over the substrate edge to form a suitable terminal post.

The hybrid power circuit represents a significant economic advantage in circuit applications, since it eliminates the necessity of making hand-wired interconnections. It also offers an economic advantage to the device manufacturer because it avoids the use of individual packages for each device. As a result, it is safe to say that much of the emphasis on packaging for medium power, high volume applications will be along these lines.

6.8 REFERENCES

1. S. K. Ghandhi, *The Theory and Practice of Microelectronics*, John Wiley & Sons, New York, 1968.
2. J. Cornu and R. Sittig, "The Influence of Doping Inhomogeneities on the Reverse Characteristics of Semiconductor Power Devices," *IEEE Trans. Electron Devices*, **ED-22**, No. 3, pp. 108–114 (1975).
3. K. V. Ravi, "The Heterogeneous Precipitation of Silicon Oxides in Silicon," *J. Electrochem. Soc.*, **121**, No. 8, pp. 1090–1098 (1974).
4. H. E. Bridges et al., *Transistor Technology*, Vol. 1, Van Nostrand, New York, 1958.
5. A. Mayer, "The Quality of Starting Silicon," *Solid State Technol.* pp. 38–45, April 1972.
6. K. V. Ravi and C. J. Varker, "Growth Striations and Swirls in Float-Zone Single Crystals," *Semiconductor Silicon*, published by the Electrochemical Society, Princeton, N.J., 1973, pp. 136–149.
7. H. A. Herrmann and H. Herzer, "Doping of Silicon by Neutron Radiation," *J.*

REFERENCES

313

Electrochem. Soc., **122**, No. 11, pp. 1568–1569 (1975).

8. H. M. Janus and O. Malmros, "Application of Thermal Neutron Irradiation for Large Scale Production of Homogeneous Phosphorus Doping of Floatzone Silicon," *IEEE Trans. Electron Devices*, **ED-23**, No. 8, pp. 797–802 (1976).

9. F. A. Trumbore, "Solid Solubility of Impurity Elements in Germanium and Silicon," *Bell Sys. Tech. J.*, **39**, pp. 205–233 (1960).

10. G. L. Vick and K. M. Whittle, "Solid Solubility and Diffusion Coefficients of Boron in Silicon," *J. Electrochem. Soc.*, **116**, No. 8, pp. 1142–1144 (1969).

11. J. E. Lawrence, "The Cooperative Diffusion Effect," *J. Appl. Phys.*, **37**, pp. 4106–4112 (1966).

12. Y. Yukimoto et al., "Effect of Tin Diffusion of Impurities in Transistor Structure," *Semiconductor Silicon*, published by the Electrochemical Society, Princeton, N.J. 1973, pp. 692–700.

13. P. Rai-Choudhury and F. A. Selim, "The Diffusion and Incorporation of Aluminum into Silicon," *Extended Abstracts*, 149th Electrochemical Society Meeting, Vol. 76–1, Washington, D.C. May 2–7, 1976.

14. J. A. Becker, "Silicon Wafer Processing by Application of Spun-on Doped and Undoped Silica Layers," *Solid State Electron.*, **17**, No. 1, pp. 87–94 (1974).

15. M. Hansen and A. Anderko, *Constitution of Binary Alloys*, McGraw-Hill Book Co., New York, 1958.

16. J. Bloem, "Trends in the Chemical Vapor Deposition of Silicon," *Semiconductor Silicon*, published by the Electrochemical Society, Princeton, N.J., 1973, pp. 180–190.

17. R. Denning and D. A. Moe, "Epitaxial π-ν n-p-n High Voltage Power Transistors," *IEEE Trans. Electron Devices*, **ED-17**, No. 9, pp. 711–716 (1970).

18. A. G. Milnes, *Deep Impurities in Semiconductors*, John Wiley & Sons, New York, 1973.

19. E. J. Mets, "Poisoning and Gettering Effects in Silicon Junctions," *J. Electrochem. Soc.*, **112**, No. 4, pp. 420–425 (1965).

20. P. M. Petroff et al., "Elimination of Process-Induced Stacking Faults by Preoxidation Gettering of Si Wafers," *J. Electrochem. Soc.*, **123**, No. 4, pp. 565–570 (1976).

21. T. W. Sigmon et al., "Ion Implantation Gettering of Gold in Silicon," *J. Electrochem. Soc.*, **123**, No. 7, pp. 1116–1117 (1976).

22. A. Goetzberger and W. Shockley, "Metal Precipitates in Silicon p-n Junctions," *J. Appl. Phys.*, **31**, pp. 1821–1824 (1960).

23. M. Nakamura and N. Oi, "A Study of Gettering Effect of Metallic Impurities in Silicon," *Japan. J. Appl. Phys.*, **7**, No. 5, pp. 512–519 (1968).

24. S. P. Murarka, "A Study of the Phosphorus Gettering of Gold in Silicon by Use of Neutron Activation Analysis," *J. Electrochem. Soc.*, **123**, No. 5, pp. 765–767 (1976).

25. R. L. Meek, et al., "Diffusion Gettering of Au and Cu in Silicon," *J. Electrochem. Soc.*, **122**, No. 6, pp. 786–796 (1975).

26. W. M. Bullis, "Properties of Gold in Silicon," *Solid State Electron.*, **9**, pp. 143–168 (1966).

27. K. P. Lisiak and A. G. Milnes, "Energy Levels and Concentrations for Platinum in Silicon," **18**, pp. 533–540 (1975).

28. J. M. Fairfield and B. V. Gokhale, "Gold as a Recombination Center in Silicon," *Solid State Electron.*, **8**, No. 8, pp. 685–691 (1965).

29. M. D. Miller, "Differences between Platinum and Gold-Doped Silicon Power Devices," *IEEE Trans. Electron Devices*, **ED-23**, No. 12, pp. 1279–1283 (1976).

30. K. P. Lisiak and A. G. Milnes, "Platinum as a Lifetime Control Deep Impurity in Silicon," *J. Appl. Phys.*, **46**, No. 12, pp. 5229–5235 (1975).

31. B. J. Baliga and S. Krishna, "Optimization of Recombination Levels and their Capture Cross Sections in Power Rectifiers and Thyristors," *Solid State Electron.*, **20**, No. 3, pp. 225–232 (1977).

32. F. Larin, *Radiation Effects in Semiconductor Devices*, John Wiley & Sons, New York, 1968.

33. P. Rai-Choudhury et al., "Electron Irradiation Induced Recombination Centers in Silicon Minority Carrier Lifetime Control," *IEEE Trans. Electron Devices*, **ED-23**, No. 8, pp. 814–818 (1976).

34. A. O. Evwaraye and B. J. Baliga, "The Dominant Recombination Centers in Electron-Irradiated Semiconductor Devices," *J. Electrochem. Soc.*, **124**, No. 6, pp. 913–916 (1977).

35. R. W. Lade and A. G. Jordon, "A Study of Ohmicity and Exclusion in High-Low Semiconductor Devices," *IEEE Trans. Electron Devices*, **ED-10**, No. 4, pp. 268–272 (1963).

36. L. R. Rice, Ed., *Silicon Controlled Rectifier Designers Handbook*, Westinghouse Electric Corp., Semiconductor Division, Youngwood, Pa., 1970.

37. D. R. Grafham and J. C. Hey, Eds., *SCR Manual*, 5th ed., Semiconductor Products Department, General Electric Co., Syracuse, N.Y. 1972.

38. S. B. Dewan and A. Straughen, *Power Semiconductor Circuits*, John Wiley & Sons, New York, 1975.

39. G. A. Lang et al., "Thermal Fatigue in Silicon Power Transistors," *IEEE Trans. Electron Devices*, **ED-17**, No. 9, pp. 787–793 (1970).

40. N. D. Zommer et al., "Reliability and Thermal Impedance Studies of Soft Soldered Power Transistors," *IEEE Trans. Electron Devices*, **ED-23**, No. 8, pp. 843–850 (1976).

41. J. DeWarga and J. E. Mungenast, "Fabrication and Application of 76 mm and 102 mm Power Semiconductors," *IEEE Conference Record*, Industry Applications Society, Pittsburgh, October 7–10, 1974, pp. 261–268.

42. B. J. Baliga and S. K. Ghandhi, "Growth of Silica and Phosphosilicate Films," *J. Appl. Phys.*, **44**, No. 3, pp. 990–994 (1973).

43. M. Conti and F. Tegnani, "Electrical Properties of Silicone Films on Silicon," *J. Electrochem. Soc.*, **116**, No. 3, pp. 377–380 (1969).

44. Y. E. Sun and J. C. Driscoll, "Direct Bonded, Glass Passivated Power Modules: The New IC's of Power Electronics," 1975 *Conference Record*, IEEE Industry Applications Society Meeting, Atlanta, Ga., September 28–October 2, 1975, pp. 50–54.

45. K. Miwa et al., "Glass Passivation of Silicon Devices by Electrophoresis," *Proc., Denki Kagaku*, **40**, No. 7, pp. 478–484 (1972).

46. R. E. Blaha and W. R. Fahrner, "Passivation of High Breakdown Voltage *p-n-p* Structures by Thermal Oxidation," *J. Electrochem. Soc.*, **123**, No. 4, pp. 515–518 (1976).

47. C. A. Neugebauer et al., "Channel Shortening in MOS Transistors During Junction Walk-out," *Appl. Phys. Lett.*, **19**, No. 8, pp. 287–289 (1971).

48. B. A. McDonald, "Avalanche Degradation of h_{FE}," *IEEE Trans. Electron Devices*, **ED-17**, No. 10, pp. 871–878 (1970).

49. T. Matsushita et al., "Highly Reliable High-Voltage Transistors by Use of the SIPOS Process," *IEEE Trans. Electron Devices*, **ED-23**, No. 8, pp. 826–830 (1976).

50. J. F. Burgess et al., "The Direct Bonding of Metals to Ceramics by the Gas-Metal Eutectic Method," *J. Electrochem. Soc.*, **122**, No. 5, pp. 688–690 (1975).

Symbols

Symbol	Description	Section
A	Cross-section area	1.4.3
a	Cross-section area	1.4.3
$\mathcal{C}$	The grade constant	2.4.4
B	Symmetry factor	3.3
BV	Breakdown voltage	2.3
BV_{CBO}	Breakdown voltage in the common base configuration	4.5.1.1
BV_{CEO}	Breakdown voltage in the common emitter configuration	4.5.1.1
BV_{PN}	Breakdown voltage of a p-n diode	2.4.2
BV_{PT}	Breakdown voltage of a punched-through diode	2.4.2
BV_{CY}	Breakdown voltage of a cylindrical junction	2.5.1
BV_S	Breakdown voltage of a spherical junction	2.5.2
b	Ratio of electron to hole mobility	1.2
C	Specific heat	1.5.2.2
C_0	Capacitance per unit area	2.5.5
C_t	Transition capacitance	3.1.2
C_t^*	Average value of transition capacitance	3.1.2
C_{tC}	Collection transition capacitance	4.4
C_{tE}	Emitter transition capacitance	4.4

Symbol	Description	Section
C_θ	Thermal capacity	6.5.1.1
D	Distance between emitter shorts	5.4.2
D	Diffusion coefficient	2.4.5
D_a	Ambipolar diffusion constant	1.2
D_n	Diffusion constant for electrons	1.2
D_{nB}	Diffusion constant for electrons in the base	3.1.1.1
D_{nP}	Diffusion constant for electrons in a p-region	3.3.1.3
D_p	Diffusion constant for holes	1.2
D_{pN}	Diffusion constant for holes in an n-region	3.3.1.3
d	Diameter of emitter shorts	5.4.2
d	Half width of a p-i-n diode	3.3
E_c	Conduction band energy level	1.1
E_f	Fermi level	1.1
E_g	Energy band gap	3.2.1
E_i	Intrinsic energy level	1.1
E_r	Activation energy of recombination centers	1.1
E_v	Valence band energy level	1.1
$\mathcal{E}$	Electric field	1.2
$\mathcal{E}^*$	Average electric field	3.3
$\mathcal{E}_C$	Electric field at the cathode	1.3
$\mathcal{E}_p$	Peak electric field	2.4.1
$\mathcal{E}_p'$	Peak electric field at breakdown	2.4.1
$\mathcal{E}_{pCY}'$	Peak electric field at breakdown in a cylindrical junction	2.5.1
$\mathcal{E}_{pPP}'$	Peak electric field at breakdown in a parallel plane junction	2.5.1
$\mathcal{E}_{pS}'$	Peak electric field at breakdown in a spherical junction	2.5.2
f_t	Gain-bandwidth product	4.4
h	Injection ratio	1.1
I_A	Anode current	5.2
I_C	Collector current	4.4
I_F	Forward current	3.1.1.1
I_G	Gate current	5.2
I_H	Holding current	5.2.2
I_K	Cathode current	5.2
I_L	Leakage current	5.2
I_R	Reverse current	3.1.2
I_x	Current at x	5.6.1
i_A	Time varying value of anode current	5.4.1

Symbol	Description	Section
i_n	Electron current	3.1.2
J	Current density	1.3
J_C	Collector current density	4.2
$J_{F,\text{diff}}$	Forward current density due to diffusion	3.1.1
$J_{F,\text{high}}$	Forward current density at high injection levels	3.1.1
$J_{F,\text{rec}}$	Forward current density due to recombination	3.1.1
J_n	Electron current density	2.1
J_{nE}	Electron current density in the emitter	4.3.1
J_{nP}	Electron current density in the p-region	3.3.1.3
J_{ns}	Electron saturation current density	3.3.1.3
J_0	Critical current density	4.3.2
J_0'	Critical current density	4.3.3
J_0''	Critical current density	4.3.3
J_p	Hole current density	2.1
J_{pC}	Hole current density in the collector	4.2
J_{pE}	Hole current density in the emitter	4.3.1
J_{pN}	Hole current density in the n-region	3.3.1.3
J_{ps}	Hole saturation current density	3.3.1.3
J_{sc}	Space charge generation current density	2.1
J_{scl}	Space charge limited current density	1.3
J_X	Current density at X	1.4.3
K	Constant	1.3
K	Thermal conductivity	1.5.2.1
K'	Damage factor	6.3.1.2
K_1	Correction factor	5.1
K_2	Correction factor	5.1
k	Boltzmann's constant	1.1
L	Length	1.3
L_a	Ambipolar diffusion length	3.3
L_n	Electron diffusion length	3.1.1
L_{nP}	Electron diffusion length in a p-region	3.3.1.3
L_p	Hole diffusion length	3.2
L_{pN}	Hole diffusion length in an n-region	3.3.1.3
L_{pN1}	Hole diffusion length in the $N1$ region	5.1
l_E	Length of the emitter	4.3.2
M	Multiplication factor	2.2
M_n	Multiplication factor for electrons	2.3
M_p	Multiplication factor for holes	2.3
M_{sc}	Multiplication factor for the space charge generation component of current	2.3

Symbol	Description	Section
N	Net impurity concentration	3.2
N_a	Acceptor doping concentration	2.4.1
N_B	Background carrier concentration	1.3
N_c	Density of states at the conduction band edge	1.1
N_d	Donor concentration	1.4.1
N_E^+	Ionized impurity concentration in the emitter	4.3.1
N_{eff}	Effective doping concentration	2.4.1
N_0	Surface concentration	2.4.5
N_r	Concentration of recombination centers	1.1
N_r^+	Concentration of positively ionized recombination centers	1.1
$\overline{N}_r^+$	Equilibrium concentration of positively ionized recombination centers	1.1
N_r^0	Concentration of neutral recombination centers	1.1
$\overline{N}_r^0$	Equilibrium concentration of neutral recombination centers	1.1
N_{ss}	Surface state density	2.6.2.1
N_v	Density of states at the valence band edge	1.1
$N(x,t)$	Impurity concentration	2.4.5
N_ν	Doping concentration in the ν-region	3.3.1
N_ν^+	Ionized donor concentration in the ν-region	3.3.1
n	Electron concentration	1.1
n^*	Average electron concentration	3.3
n'	Excess electron concentration	1.1
n'	Mobile electron concentration in a depletion layer	4.3.3
$\overline{n}$	Equilibrium electron concentration	1.1
n_i	Intrinsic electron concentration	1.5.1
n_{i0}	Intrinsic carrier concentration in the absence of energy gap narrowing	3.2.1
$\overline{n}_P$	Equilibrium concentration of electrons in a p-region	2.1
$n_P(0)$	Electron concentration in a p-region at the edge of the depletion layer	3.1.1
n_r	Density of empty recombination centers	1.3.1
P	Power	1.5.2.2
P_1	Power dissipation	4.5.2
P_2	Power dissipation	4.5.2
p	Hole concentration	1.6

Symbol	Description	Section
p^*	Average hole concentration	3.3
p'	Excess hole concentration	1.1
$\bar{p}$	Equilibrium hole concentration	1.1
$\bar{p}_N$	Equilibrium concentration of holes in a p-region	2.1
p_P	Hole concentration in a p-region	3.1.1
$p_P(0)$	Hole concentration in a p-region, at the edge of a depletion layer	3.1.1
Q	Heat per unit area	1.5.2.1
Q_B	Base charge	3.1.1.1
Q_E	Emitter charge	4.3.1
Q_L	Charge on left side	3.3.2
Q_{N1}	Charge in the $N1$ region	5.1
Q_{N2}	Charge in the $N2$ region	5.1
Q_{P2}	Charge in the $P2$ region	5.1
Q_R	Charge on right side	3.3.2
Q_S	Stored charge	4.3.2
$Q(t)$	Time dependent stored charge	3.1.2
q	Magnitude of charge on the electron	1.3
R	Resistance	3.1.2
R	Recombination rate	3.3
R_{cn}	Capture rate for electrons	1.1
$\bar{R}_{cn}$	Equilibrium value of the electron capture rate	1.1
R_{cp}	Capture rate for holes	1.1
$\bar{R}_{cp}$	Equilibrium value of the hole capture rate	1.1
R_{en}	Emission rate for electrons	1.1
$\bar{R}_{en}$	Equilibrium value of the electron emission rate	1.1
R_{ep}	Emission rate for holes	1.1
$\bar{R}_{ep}$	Equilibrium value of the hole emission rate	1.1
R_{P2}	Total resistance of the $P2$ base, measured between gate contacts	5.6.3
R_θ	Thermal resistance	6.5.1
$R_{\theta CA}$	Thermal resistance, case to ambient	6.5.1
$R_{\theta DC}$	Thermal resistance, die bond to case	6.5.1
$R_{\theta S\text{-air}}$	Thermal resistance, heat sink to air	Table 6.5
$R_{\theta S\text{-water}}$	Thermal resistance, heat sink to water	Table 6.5
$R_{\theta JA}$	Thermal resistance, junction to ambient	6.5.1
$R_{\theta JD}$	Thermal resistance, junction to die bond	6.5.1
$R_{\theta JS}$	Thermal resistance, junction to heat sink	Table 6.5
$R_{\theta t}$	Transient thermal impedance	6.5.1.1

Symbol	Description	Section
$R_{\theta 1}$	Thermal resistance	4.5.2
$R_{\theta 2}$	Thermal resistance	4.5.2
r	Radius	2.5.1
r_d	Radius of edge of a depletion layer	2.5.1
r_j	Radius of curvature of a junction	2.5.1
r_0	Radius of a mesoplasma core	1.5.2.2
r_0	Gate radius	5.4.1
S_m	Surface of maximum potential	1.5.2.1
s	Symmetry factor	3.3.1.3
T	Temperature in degrees Kelvin	1.1
T_A	Ambient temperature	1.5.2.2
T_M	Core temperature	1.5.2.2
t	Time	1.1
t_B	Base transit time	3.1.1.1
t_d	Delay time	5.4
t_f	Fall time	3.1.2
t_{N1}	Transit time in $N1$ region	5.4
t_{OL}	Time to initiate a space charge region on the left side	3.3.2
t_{OR}	Time to initiate a space charge region on the right side	3.3.2
t_{P2}	Transit time in $P2$ region	5.4
t_r	Rise time	5.4
t_s	Storage time	3.1.2
U	Generation rate	1.1
$U_{\max}$	Maximum generation rate	3.1.1
u_s	Spreading velocity	5.4.1
V	Voltage	1.3
V_A	Applied voltage	1.3
V_A'	Voltage across the depletion layer	3.1.1
V_{AK}	Voltage drop, anode to cathode	5.2
V_{BE}	Base-emitter voltage	4.5.2
V_{BF}	Forward blocking voltage	5.2
V_{BR}	Reverse blocking voltage	5.1
V_{CB}	Collector-base voltage	4.3.3
V_{CC}	Collector supply voltage	4.2.1
V_{CE}	Collector-emitter voltage	4.2.1
V_D	Forward voltage drop across a diode	3.1.2
V_F	Forward voltage	3.1.2

Symbol	Description	Section
V_{FP}	Voltage on a field plate	2.5.4
V_{J3}	Forward voltage across the $J3$ junction	5.2.2
V_j	Junction voltage	2.3
V_L	Voltage drop across the left hand side junction	3.3.1
$V_{L,sc}$	Space charge voltage on the left hand side junction	3.3.2
V_M	Voltage drop across the mid-region	3.3
V'_M	Modified voltage drop across the mid-region	3.3.1.3
V_0	Voltage drop between shorts	5.4.2
V_{REV}	Reverse voltage	3.1.2
$V_R`$	Voltage drop across the right hand side junction of a p-i-n diode	3.3.1
$V_{R,sc}$	Space charge voltage across the right hand side junction	3.3.2
V_T	Threshold voltage	1.4
$V(x,t)$	Voltage as a function of distance and time	2.5.5
$V_\mathscr{E}$	Voltage drop due to an electric field	3.1.1
v	Time varying value of voltage	3.1.2
v_{AK}	Time varying value of anode to cathode voltage	5.4.1
$v_{\lim}$	Limiting velocity	1.4.1
v_{th}	Electron thermal velocity	2.7
W	Width	2.1
W'	Width of depletion layer at breakdown	2.2
W_B	Base width	3.1.1.1
W'_B	Effective base width	4.3.2
W_{CIB}	Current induced basewidth	4.3.3
W_G	Overlap of a cathode short	5.2.2
W_H	Depletion layer width on the heavily doped side	2.6.2
W_K	Width of a cathode short	5.2.2
W_L	Depletion layer width on lightly doped side	2.6.2
W_N	Width of an n-region	2.1
W_P	Width of a p-region	2.1
W_ν	Width of the ν-region	4.2
w_E	Width of the emitter	4.3.2
x	Distance	1.3
x_B	Depletion layer width in the presence of current flow	4.3.3

Symbol	Description	Section
x_{B0}	Depletion layer width for no current flow	4.3.3
x_C	Collector depletion layer width	4.4
x_0	Effective injection width	4.3.2
x_0	Space constant	5.6.1
Z	Depth of an emitter short	5.2.2
α	Common base current gain	4.3
α_i	Ionization coefficient	2.3
$\tilde{\alpha}$	Small signal alpha	5.3.1
α_n	Ionization coefficient for electrons	2.2
α_{npn}	Common base current gain of the npn transistor	5.2
α_p	Ionization coefficient for holes	2.2
α_{pnp}	Common base current gain of the pnp transistor	5.2
α_0	Low frequency common base current gain	4.3
α_T	Base transport factor	4.3
α_{Te}	Common emitter base transport factor	4.3
$(\alpha_T)_{npn}$	Base transport factor of a npn transistor	5.2.2
$(\alpha_T)_{pnp}$	Base transport factor of the pnp transistor	5.2
β_0	Low frequency common emitter current gain	4.2
β_{off}	Turnoff current gain	5.6
β_{pnp}	Common emitter current gain of the pnp transistor	5.1.1
Γ	The gamma function	2.4.4
γ	Injection efficiency	4.3
γ_e	Common emitter injection efficiency	4.3
$\gamma_{e,\text{low}}$	Low level injection efficiency	4.3.2
γ_1	Phonon-assisted recombination constant	1.1
γ_3	Auger recombination constant	1.1
γ_{3n}	Auger recombination constant in p^+-material	1.1
γ_{3p}	Auger recombination constant in n^+-material	1.1
δ	Minimum turnoff width	5.6.1
ε	Relative permittivity	1.3
ε_0	Permittivity of free space	1.3
η	Ratio of diode width to full depletion layer width	2.4.2
θ	Temperature	1.5.2.1
θ	Bevel angle	2.6.2
θ_i	Intrinsic temperature	1.5.1
θ_m	Maximum temperature	1.5.2.1

Symbol	Description	Section
λ	A space constant	2.4.5
μ_n	Electron mobility	1.2
μ_p	Hole mobility	1.2
μ_ν	Mobility in the ν-region	4.3.3
ρ	Space charge density	1.3
ρ	Electrical resistivity	1.5.2.1
ρ_L	Space charge density at the left side	3.3.2
$\rho_{P2\square}$	Sheet resistance of the $P2$ region	5.2.2
$\rho_\square$	Sheet resistance of an oxide surface	2.5.5
σ_B	Base conductivity	4.3.2
σ_{N1}	Conductivity of the $N1$ region	5.4
σ_{P2}	Conductivity of the $P2$ region	5.4
σ_p	Hole capture cross section	2.7
τ_A	Auger lifetime	4.3.1
τ_a	Ambipolar lifetime	1.2
τ_D	Initiation time delay	1.5.2.2
τ_{eff}	Effective lifetime	1.1
τ_F	Effective lifetime in the forward direction	3.1.2
τ_{high}	High level lifetime	3.3.2
τ_{low}	Low level lifetime	3.3.2
τ_n	Electron lifetime	1.4
$\tau_{n,\text{low}}$	Low level electron lifetime	1.4
τ_{n0}	Electron lifetime, extrinsic case	1.1
τ_0	Minority carrier lifetime before radiation	6.3.1.2
τ_p	Minority carrier lifetime (holes)	1.1
$\tau_{p,\text{high}}$	High level hole lifetime	1.1
$\tau_{p,\text{low}}$	Low level hole lifetime	1.1
τ_{p0}	Hole lifetime, extrinsic case	1.1
τ_R	Effective reverse lifetime	3.1.2
τ_{sc}	Space charge generation lifetime	1.1
τ_ν	Lifetime in the ν-region	4.2
ϕ	Potential	1.5.2.1
ϕ	Bevel parameter	2.6.2
ϕ_e	Radiation fluence density	6.3.1.2
ϕ_m	Maximum potential	1.5.2.1
ψ	Contact potential	4.3.3

INDEX